Frozen Oceans

THE FLOATING WORLD OF PACK ICE

DAVID N. THOMAS

FIREFLY BOOKS

A FIREFLY BOOK

To Rachel & Naomi

Published by Firefly Books Ltd. 2004

First printing

Publisher Cataloging-in-Publication Data (U.S.)

Thomas, David N.
    Frozen oceans : the floating world of pack ice / David N. Thomas. — 1st ed.
[224] p. : col. ill., photos., maps.;   cm.
Includes bibliographical references and index.
Summary: Detailed and scientific background to the ecosystem of pack ice,
its living communities and the structure of the ice.
ISBN 1-55407-000-7
1. Ecology — Polar regions.   2. Marine biology — Polar regions. 3. Polar regions. I. Title.
577.7 22   QH541.5.P6.T56 2004

National Library of Canada Cataloguing in Publication

Thomas, David N. (David Neville), 1962-
      Frozen oceans : the floating world of pack ice / David N. Thomas.
Includes bibliographical references and index.
ISBN 1-55407-000-7
    1. Sea ice. I. Title.
GB2403.2.T48 2004   551.34'3   C2004-902727-1

Published in the United States in 2004 by
Firefly Books (U.S.) Inc.
P.O. Box 1338, Ellicott Station
Buffalo, New York 14205

Published in Canada in 2004 by
Firefly Books Ltd.
66 Leek Crescent
Richmond Hill, Ontario L4B 1H1

Edited by Celia Coyne
Index by Angie Hipkin
Designed by Mercer Design

Printed in Singapore

Front cover Water movements force the grease ice crystals to coagulate into ice pancakes.

Back cover (left to right) Setting up an ice camp with the local wildlife.
Specially-strengthened ice-breaking ships are central to pack ice research.
Large pressure ridges tower over a scientist in the Baltic Sea pack ice.

Contents

Preface

PACK ICE IS A LAYER OF FROZEN SEAWATER typically seen floating on the polar oceans, although it is also a feature of seas such as the Baltic, Caspian and White. It varies in thickness from a few centimetres to tens of metres and at its maximum extent covers up to 13% of the Earth's surface. This makes pack ice one of the major habitat types on the planet, similar, in terms of surface area, to deserts and tundra. Ever since the sealers and whalers began to navigate through regions of pack ice at the end of the 19th century, the physical and mechanical properties of sea ice have been a focus of intense research.

Pack ice, or as it is commonly called sea ice, not only dominates polar regions but it is also central to global ocean circulation as well as global climate patterns. On a smaller scale, the formation, consolidation and subsequent melt of millions of square kilometres of ice have a fundamental impact on every ecosystem in which sea ice forms. A plethora of micro-organisms live within the ice itself, and sea ice is an important ephemeral feature affecting the seasonal dynamics of many plankton species, as well as the mammals and birds that depend on plankton as food. Some of the larger animals affected by pack ice include whales, seals, polar bears, walruses and even Arctic foxes.

Although the microscopic ice dwellers were discovered over 150 years ago, it is only in the past 25 years that the biology and chemistry of sea ice have become the focus of systematic investigations. In recent years the study of this extreme environment has intensified, fuelled by the biotechnological potential of micro-organisms from cold habitats. Astrobiologists have even explored their potential as proxies for life on an early Earth or extraterrestrial systems such as Mars and the moons of Jupiter. This book is an introduction to the large-scale distribution of pack ice around the world. The physics of ice formation, ocean dynamics and the consequences for the biology that depends on sea ice is discussed. Historical aspects of pack ice exploration are introduced together with a synopsis of many of the modern-day techniques used to study sea ice.

Places referred to in the text can be found on the maps on pp. 210 and 211.

Author

DAVID THOMAS IS A READER IN MARINE BIOGEOCHEMISTRY and heads the biogeochemistry group in the School of Ocean Sciences at the University of Wales, Bangor. He has worked with sea-ice related issues since 1991 when he made his first research expedition to the pack ice of the Weddell Sea, Antarctica. He has participated in further sea-ice expeditions in the Southern Ocean, Arctic, Baltic Sea and the White Sea, Russia. He was awarded his PhD from the Botany Department, University of Liverpool in 1988. Before taking up his position in Bangor in 1996, he held four research positions in Germany at the Universities of Bremen and Oldenburg, the Alfred Wegener Institute for Marine and Polar Research, Bremerhaven and the Centre for Marine Tropical Ecology, Bremen.

Left
The author at Drescher Inlet ice camp in the Weddell Sea, Antarctica.

What is pack ice?

1

"NOW WE ARE IN THE VERY MIDST OF WHAT THE PROPHETS WOULD HAVE HAD US DREAD SO MUCH. THE ICE IS PRESSING AND PACKING ROUND US WITH A NOISE LIKE THUNDER. IT IS PILING ITSELF UP INTO LONG WALLS, AND HEAPS HIGH ENOUGH TO REACH A GOOD WAY UP THE FRAM'S RIGGING; IN FACT, IT IS TRYING ITS UTMOST TO GRIND THE FRAM INTO POWDER."

THIS FRIGHTENING DESCRIPTION of the awesome power of the ice was written by the Norwegian Fridtjof Nansen on Friday, October 13, 1893, as his robust ship, the *Fram*, became stuck in the Arctic pack ice. Together with 12 other adventurers, he was at the beginning of an epic three-year voyage to traverse the Arctic Ocean. During the journey they would be exposed to conditions that would test their seamanship and survival skills to the absolute limits. Nansen, arguably the greatest polar explorer, knew that there was nothing he or his crew could do to combat the extreme physical forces that dominate moving fields of pack ice. He recognised that their survival was down to the strength of their ship and having enough provisions to last the long frozen winter months.

This can be taken as the point at which scientific investigation of pack ice opened up and understanding of its complexity, and its role in the functioning of the global environment, began to develop and useful practical information in navigation, fishing and climate forecasting was made available. Before this the 'civilized' world had mostly looked on pack ice as a nuisance, unpredictable, obstructive, unproductive and a potential destroyer and wrecker. But at the same time in the Arctic but not in the Antarctic, there was a race of people, Eskimos or, more properly, Inuit, who had knowledge and deep understanding of pack ice. They had lived with it, travelled on it and hunted their food from it for thousands of years. Mostly these people with their natural wisdom were ignored until comparatively recently. More will be said about them later.

Opposite
Winter in the pack ice is a hostile place to work with temperatures down to −40°C (−40°F), and permanent gloom.

Even with all the developments in shipbuilding that have produced immensely strong metal-hulled ice-breaking ships crammed full of the latest satellite-based navigation aids, pack ice just a few centimetres thick can hinder modern day seafarers just as much as the early polar explorers in their wooden vessels. The pack-ice regions of the world are hostile, and only inhabited by man at the very fringes of their extent. They are places of extreme low temperatures, darkness for long periods of the year and severe winds.

My first impression of the Antarctic pack ice was actually one of extreme beauty. The tranquillity I felt was a complete contrast to the buffeting ocean that had been my lot as the research ship travelled 15 days from Cape Town in South Africa towards the Southern Ocean. But a closer inspection of the huge ice floes, up to several metres thick, effortlessly moving on the ocean surface, rafting on top of each other and forming massive ridges of ice blocks twice my height, was a humbling experience that inspired a healthy caution. The pack ice is not a silent place, and when a ship slowly negotiates a passage between ice floes there is a constant creaking and grinding of ice, almost a haunting groaning that emphasises the alien nature of the frozen landscape.

Below
Icebergs are sculpted into spectacular shapes as they gradually melt in the open ocean.

Icebergs are not pack ice

Pack ice is mostly frozen seawater. In simple terms, the surface of the ocean is cooled down and ice crystals form. These crystals rise to the surface and coalesce to form a frozen layer on the surface of the water. This layer can become thicker, break open and refreeze. Slabs of ice can raft on top of each other and deform, as in Nansen's description above. Ice formed from seawater, or sea ice, therefore varies from loose aggregations of ice crystals through to structures that are tens of metres thick.

However, when most people think of the polar oceans and seas they picture huge floating icebergs that have broken off from coastal glaciers or ice shelves. There is no doubt that icebergs are a distinctive feature of many pack-ice regions, but they are produced from outside of the pack ice, and are only a small part of a much larger frozen seawater system. Icebergs are not made in the sea, but are large chunks of freshwater ice that have been built up over thousands of years by the gradual freezing of snow and ice on glaciers covering land.

Nevertheless, one of the most striking sights for anyone who ventures into polar waters is the wonderful array of iceberg shapes. Wind and wave action combine to create bizarre forms, from cathedral-like spires and pinnacles, to caverns and other improbable structures. Icebergs also come in a spectacular range of colours, from the familiar blues and whites through to dark green and even black.

Ice shelves and glaciers

Many of the 'ice stories' that hit the news headlines are in fact talking about icebergs and the breaking up of ice shelves (the thick plates of ice, fed by glaciers, that float on the ocean around much of the Antarctic continent). One of the most dramatic ice-shelf events in recent years was the collapse and break up of the Larsen B ice shelf on the eastern side of the Antarctic Peninsula in 2002. Within just over a month 3250 km² (1255 sq miles) of the shelf broke off from the continent and disintegrated into thousands of icebergs that drifted off into the Weddell Sea. The shelf was estimated to be in the order of 12,000 years old, and was 220 m (722 ft) thick in places. The amount of ice released in this short time was in the order of 720 billion tonnes. The largest iceberg to calve in recent years was iceberg B-15, which calved from the Ross Ice Shelf in 2000. It was a massive 295 km (183 miles) long and up to 25 km (16 miles) wide, a total area of 11,000 km² (4247 sq miles). Since 1974 it is estimated that seven ice shelves around the Antarctic Peninsula have shrunk by a total of about 13,500 km² (5212 sq miles).

Icebergs are not just a feature of the Antarctic, although there are far fewer icebergs in the Arctic because of the less extensive ice shelves and large glaciers. Most of the icebergs in the north Atlantic come from about 100 iceberg-

Below
During the period January to March 2002, the Larsen B ice shelf broke up into thousands of smaller icebergs. The extent of the shelf in 2002 was already very much less than in 1995.

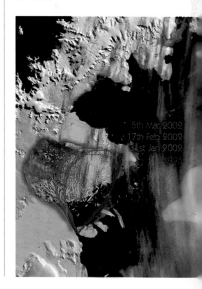

producing glaciers along the coast of Greenland, while a few originate in the eastern Canadian Arctic islands. The Jakobshavn glacier on the western shore of Greenland, which flows up to 20 m (66 ft) per day, produces 10% of the Greenland icebergs (approximately 1350 annually). The annual yield of icebergs in the Arctic is estimated to be up to 40,000 medium to large icebergs, although only 1–2% or so of these are of sufficient size to reach the open ocean intact.

What happens to icebergs?

Once in the ocean the fate of the iceberg depends on many factors. In shallow waters icebergs can ground on the seabed, sometimes for several years. Collections of grounded icebergs can reach such numbers that they are known as iceberg graveyards. Some 90% of an iceberg is actually under the water line, and this mass of ice forms a huge keel that is caught by surface currents. The upper part of a berg, which can tower to over 100 m (328 ft), 'catches' the wind very effectively and the passage of an iceberg is therefore determined by a combination of winds, waves and currents. Bergs can move anywhere up to 1 m (3ft) per second and are estimated to travel up to 40 km (25 miles) per day. It is intriguing that pack ice may drift in one direction, propelled by the wind, whereas icebergs may move in a quite different direction propelled by deep-water currents.

In fact, oceanographers track icebergs by satellites to gain valuable information about the movement of surface waters. Some of the biggest bergs, with areas of several thousand square kilometres, last for many years before eventually melting, and are therefore very useful tools for measuring long-term trends in the movement of surface waters. Iceberg B-9, which broke free from the Ross Ice Shelf in 1987, split into two smaller icebergs, B-9A and B-9B, which were still being tracked half-way round the Antarctic continent 13 years later.

As icebergs melt they may calve off smaller bergs, eventually disintegrating into even smaller bergs. Categories of iceberg sizes range from very large (greater than 10 million tonnes and hundreds of square kilometres in surface area) to large, medium and small bergs. Then there are 'bergy bits' and 'growlers', which are the size of a grand piano. The rate of melting depends on the water temperature and the amount of wave action; a berg in water around 0°C (32°F) may deteriorate at a rate 10% slower than the same berg in 10°C (50°F) waters.

In Arctic waters icebergs can prove a hazard to the offshore drilling platforms, and there are programmes to monitor iceberg production and trajectories by various groups including coastguard agencies. In certain circumstances the towing of icebergs to deflect them from their path is the only course of action remaining to protect offshore structures.

Opposite top
In shallow coastal areas iceberg graveyards can form when icebergs become stuck on the seabed below, sometimes for many years.

Opposite bottom
Eventually icebergs disintegrate to form ever-decreasing lumps of ice.

Icebergs in the pack ice and as a freshwater source

The massive icebergs clearly also affect the distribution patterns of the pack ice, sometimes on a massive scale of many hundreds of square kilometres. Over a 10-year period a massive floating glacier, the Ninnis glacier tongue in east Antarctica, has gradually broken up to produce hundreds of icebergs, the largest of which had a surface area of around 800 km^2 (309 sq miles). Bergs such as these, especially if grounded, affect the strength and direction of ocean and pack-ice flow, forming a huge physical barrier akin to an island. The break up of the Ninnis glacier and the numerous icebergs it has produced over the last 10 years have been closely monitored by sea-ice researchers using satellite monitoring techniques. They report clear links between profound changes in average sea-ice thickness, density and distribution with the numbers and movements of these icebergs.

Since icebergs are made of freshwater several schemes have suggested towing icebergs to warmer climes to use the meltwater as a freshwater source. This is not as far-fetched as it may at first seem. There are colossal water reserves locked up in the polar icecaps, and the water contained within the icebergs could go a long way to combating the severe water shortages that many people in the world suffer. The largest northern hemisphere iceberg on record was encountered near Baffin Island in 1882, with an estimated mass in excess of 9 billion tonnes of water. The Antarctic icebergs contain orders of magnitude more, the Ross Sea icebergs B-9 and B-15 are estimated to contain trillions of tonnes of water each. At time of writing the population of the world is around 6.4 billion. Therefore, just one of these giant icebergs would be enough to meet the whole of the world's water needs for a considerable time. It seems that in the future, ideas for harnessing this freshwater source will increase. On a much smaller scale, ice from icebergs in Newfoundland is collected for both bottled water and vodka production.

Sea ice is just frozen water, isn't it?

Despite the colossal amounts of water contained within floating icebergs, the numbers pale into insignificance compared to the huge areas covered by frozen seawater, the pack ice. In the Southern Ocean alone, up to 20 million km^2 (8 million sq miles) are covered by sea ice. This is an area as big as North America and Canada combined. The estimated average thickness is around 1 m (3 ft), and so there are approximately 20,000 km^3 (4800 cubic miles) of sea ice. To put this in perspective, a household chest freezer has a volume of approximately 0.25 m^3 (9 ft^3). The total world human population is around 6.5 x 10^9, and so to contain all of the Antarctic sea ice, each person on Earth would need roughly 12,300 chest freezers.

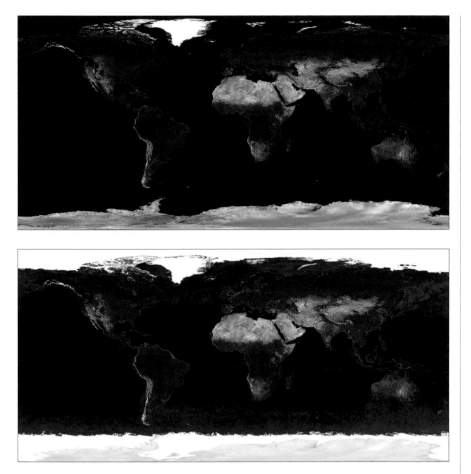

Left
The top image shows the
Arctic and Antarctic without
sea ice, whereas in the bottom
image the pack ice (during the
Antarctic winter) in both
oceans is shown.

If all of the ice presently on the Earth's landmasses (most of it lying over the Antarctic continent) melted, the sea level would rise by about 80 m (262 ft), flooding huge areas of coastal regions. Since sea ice is less dense than seawater, when it floats on the ocean surface it displaces only its own weight of seawater, so if all the sea ice melted the meltwater would simply replace the volume previously occupied by the ice. That is why there are no sea-level changes with the annual formation and melting of the millions of square kilometres of sea ice each year.

Pack ice has many names

Frozen seawater at first may be considered to be simply ice. However, it is not quite so easy, and even a cursory glance into the literature dealing with sea ice turns up a bewildering collection of ice-terms: frazil ice, slush ice, rafted ice, columnar ice, granular ice, grease ice, ice floe, porridge ice, bottom ice, infiltration ice, nilas ice, ice ridges, pancake ice, consolidated ice, pack ice, fast ice, multiyear ice, rotten ice, brown ice, anchor ice, platelet ice, ice brine, ice flowers, snow ice, superimposed ice, grey ice, black ice, white ice … to name just a few.

Many of these terms will be introduced in this book, although this collection of names pales into insignificance when compared to the range of words the Inuit have to describe the state of sea ice. Studies of the Inuit knowledge of sea ice reveal a profound understanding of the relationships between ice, water currents, tides and winds. This understanding is reflected in a rich vocabulary to describe ice-related features. In studies conducted by the Hudson Bay Programme, Inuit and Cree communities had over 80 terms relating to sea ice. Inuit hunters often link astronomical and biological observations to their own experience to exploit moving ice fields. For example, Inuit hunters have been known to predict a wind or tide change by observing mammals: in the spring a reduction in the numbers of basking seals indicates that the water is no longer freezing, since the seals will stop going onto moving ice when the thinning ice cannot support their weight.

The Inuit of Iglooik in Nanuvut, in the eastern Canadian Arctic, have an extensive ice vocabulary. For example, *ivuniraarjuruluk* is an outstanding ice formation that is used for navigation and to obtain freshwater; *uiguaq* is a thin

Below
Ice attached to the land is referred to as land-fast ice, or frequently just fast ice.

layer of newly-formed ice that attaches to the edge of an ice floe; and *ukkuartinniq* is a strip of ice that runs from the moving ice to the edge of the land-fast ice. The links between animals and ice are also intriguing: *sikutuqqijuq* is floating young ice where walrus are usually found, or the verb *tijjatuliqiyug* describes hunting walrus from the solid ice. There are, of course, many hunting terms such as *mauliq* or hunting seals through snow-covered breathing holes. Even the noise of sea ice is used by some Inuit to describe ice, such as *ivaluktaktok*, a word describing the noise of piling ice.

Sea ice that isn't pack ice

Some frozen seawater is attached to the land, floating ice shelves, or even large icebergs. This land-fast ice is not subject to the same processes as the pack ice, which is free floating and moved by wind and water currents. Land-fast ice is virtually a static structure, so the development and form of the two types of ice are quite different. This book is mostly concerned with pack ice, although as far as the frozen oceans are concerned, land-fast ice is of great consequence. Therefore, throughout the book mention will be made of land-fast ice and, where pertinent, the differences between land-fast ice and pack ice highlighted.

Land-fast ice is important as a transition between the land and the open ocean pack ice, since its relative stability is a refuge for migrating, hunting and breeding mammals, and even humans who may travel on it. It is also difficult to separate the two types of ice, since land-fast ice can break from its anchor point and drift away to form an integral part of the pack ice. Often land-fast ice is cracked and deformed by the tidal rise and fall of the underlying water. This can result in tidal cracks within the ice that are important regular openings for marine mammals and birds.

What happens when seawater freezes?

2

ONE LITRE OF SEAWATER TYPICALLY CONTAINS ABOUT 34 G
OF DISSOLVED SALTS AND IONS (MOSTLY SODIUM, CHLORIDE,
SULPHATE, MAGNESIUM, CALCIUM AND POTASSIUM). THE TERM
USED TO QUANTIFY THE SALTINESS OF SEAWATER IS SALINITY,
AND THE SEAWATER DESCRIBED HERE WOULD BE SAID TO HAVE
A SALINITY OF 34. FRESHWATER HAS A SALINITY OF 0, AND
FREEZES AT 0°C (32°F).

HOWEVER THE HIGH SALT CONTENT of seawater means that it does not begin to freeze until the temperature drops below −1.86°C (28.65°F). At this temperature, ice crystals begin to form and rise to the surface of the water body. Known as frazil ice, these crystals vary in shape, from plates to needles less than a millimetre thick, and may be up to a centimetre (half an inch) long. The surface of the ocean is a turbulent environment and water is mixed by wind and ocean circulation through tens to hundreds of metres. This mixing produces a homogeneous layer of water, so that frazil ice crystals are not limited to the very surface waters, but are formed in a much larger body of water.

Grease and pancakes

In the space of hours to days, slicks of frazil crystals form on the water surface as they are swept together by wind and water motion. These look remarkably like oil slicks and because of this and their thick soup-like consistency they are called grease ice. Although the layers tend to be just a few centimetres thick they can reach thicknesses of up to 1 m (3 ft).

In turbulent waters, the ice crystals accumulate after a few hours of further freezing to form loosely aggregated discs known as ice pancakes. At first they are only 5–10 cm (2–4 in) in diameter, but grow larger by the accumulation of more frazil ice crystals to form 'super pancakes', which are 10–50 cm (4–20 in) thick and up to several metres across. If the pancakes are formed in open water, they are a regular round shape. However, if they form along a shoreline or physical structure such as an iceberg, they elongate with the long axis parallel

Opposite
Strong, freezing winds sweep over the water, rapidly cooling down the water surface.

Top left
When seawater cools below −1.86°C (28.65°F), ice crystals form that rise to the surface collecting in wind-driven slicks of grease ice.

Top right
In a short time the grease ice (right of image) is transformed into small pancakes of ice (left of image) due to the motion of the sea surface.

Above left and right
Water movements force the grease ice crystals to coagulate into ice pancakes that start off a few centimetres in diameter growing to over several metres.

to the solid barrier. Wind and wave action raft the pancakes together, and often several end up lying on top of one another. They freeze together and after one or two days a closed ice cover (consolidated ice cover) has formed. By this stage the ice is strong enough to support the weight of heavy sledges, generators and ice-coring equipment.

Ice floe is a term that is often used to describe a piece of sea ice. This rather imprecise term is used to describe any continuous stretch of sea ice covering an area from a few square metres to more than 100 km² (40 sq miles). When large floes break-up they form a series of smaller floes, which illustrates the ambiguity of talking about an ice floe without giving an idea of its extent.

Nilas ice
Pancake ice is characteristic of turbulent waters. Sea ice that forms under calm conditions has a very different appearance. Rather than forming pancakes, the frazil ice forms into uniform sheets. These sheets are similar in appearance to the ice that forms over a freezing pond or lake. This ice is called nilas ice and

when it is less than 5 cm (2 in) thick it appears dark (dark nilas), because of the underlying water. As it thickens with age it becomes greyer and hence the terms grey or white nilas are a reflection of the age of the nilas ice.

Dark nilas has the consistency of butter or ice cream, and one of the most spectacular sights when passing through a large expanse of young nilas ice is to see the ice finger-raft. This is when sheets of nilas ice break up and move against each other in a characteristic fashion that resembles fingers interlocking. The movement of the 'fingers' is quite breathtaking as they effortlessly glide to form the complex interlocked pattern.

Columnar ice

As temperatures decrease, both consolidated pancake ice and nilas ice thicken, not necessarily by the accumulation of more frazil ice crystals, but by the quiescent growth of columnar ice. This ice is made up of vertically elongated crystals that can reach diameters of several centimetres and lengths of tens of centimetres. As they grow they add layer upon layer of ice to the underside of

Above
Thin sheets of nilas ice cut into each other in a characteristic finger-rafting.

Above
A vertical section of ice viewed under polarised light, showing regular long columnar crystals.

Above right
A vertical section of ice viewed under polarised light, showing random orientation of frazil ice crystals.

the frazil surface ice. There is normally a transition phase between the upper frazil ice layers and the columnar growth, which is crudely a mixture of frazil ice and columnar ice.

The growth of columnar ice is very much slower than the growth of frazil ice, and is greatest in less turbulent waters. The proportion of frazil ice to columnar ice in any one ice field depends largely on the turbulence of the water in which it formed. The more turbulent the water, the higher the proportion of frazil ice. This can be illustrated by comparing the composition of Arctic and Antarctic pack ice. In the Arctic 60–80% of the pack ice is columnar ice, while in the more turbulent Southern Ocean frazil ice comprises 60–80% of the total ice of most regions.

Ice that forms in deeper water

Dense clouds of ice platelets are a common feature under land-fast ice in many places around the Antarctic continent. Ice platelets are discs of ice formed when seawater flowing underneath floating ice shelves is supercooled. The platelets form and grow in water depths greater than 200 m (656 ft). The ice platelets, which can be up to 15 cm (6 in) in diameter and 3 mm thick, then rise up through the water towards the surface. These loose platelets can accumulate under any overlying sea ice, trapping seawater between the plates. Sometimes large 'platelet clouds' up to 20 m (66 ft) thick form under the ice, although generally the platelet layers are 1–5 m (3–16 ft) thick. With time a proportion of platelets become frozen into the underside of the overlying ice.

Because platelet ice is formed under floating ice shelves, it is mainly found beneath ice that is attached to land-fast ice. However, there have been records of dense accumulations of platelet ice underlying pack ice far away from ice shelves.

Anchor ice is another type of plate-like ice that forms at depths of up to 30 m (98 ft). This ice attaches itself to objects that are not frozen, and sheets of the ice often form around submerged rocks, gravel and even animals. It is a major physical disturbance in shallow waters and can entrap large benthic organisms including fish, as well as seaweeds, rocks and sediments. Anchor ice is buoyant when detached from its attachment site. It rises from the sea floor, or structure on which it formed, carrying entrapped organisms with it to the surface. Masses of seaweeds and animals weighing up to 25 kg (55 lb) have been observed under (and even incorporated into) overlying sea ice, evidently carried there by the anchor ice.

Top
Loose accumulations of platelet ice can form under Antarctic land-fast ice.

Above
An individual disc of ice or ice platelet.

Snow and ice

Sea ice is typically covered by snow, which is a very good reflector of light that would normally warm the ice. Therefore, the absorption of light, and its potential to melt ice, depends on how much snow there is. Snow, on the other hand, is also a very good insulator and ice with a thick snow cover grows much more slowly than ice covered by a thin layer of snow. In the Arctic the snow cover on sea ice is generally low – less than 30 cm (12 in) – whereas Antarctic ice is covered by an average 50 cm (20 in) of snow. This is one of the reasons why Arctic ice is on average thicker than Antarctic ice.

Snow is also heavy and can push the ice floes under the water so that seawater floods the ice surface. Refreezing of flooded snow (seawater and snow mixed together), or simply the refreezing of molten snow, can result in the production of snow ice on the very top of the ice floes. In the Antarctic, snow ice is thought to contribute up to 50% of the total pack-ice mass in some regions, although the fraction contributed solely by snow is generally less than 20%.

How does ice grow?

When ice forms on the surface of the water an immediate temperature gradient is set up across the ice layer. Thickening of the ice, or rather growth of the ice, depends on the temperature of the air at the upper ice surface and the temperature of the water beneath the ice. Ice will only continue to grow if the seawater below the ice is at or below the freezing point of seawater. Ice growth takes place at this ice-water interface and the growing ice crystals form a region called the skeletal ice layer. This is extremely fragile, and hard to sample intact. Skeletal layers vary in appearance depending on the water movements under the ice and the speed at which the ice is growing. Generally these growing edges comprise blades or plates of ice aligned vertically from the underside of the overlying ice floe. As more ice is added to the lowermost growing edges of the blades, the ice consolidates by forming cross walls to form a more solid structure.

Theoretically, ice could continue growing and growing. This is clearly not the case and the combination of the heat of the ocean, snow cover and air temperature interact to limit the growth of sea ice by ice crystal growth alone to around 1 m (3 ft) in the Antarctic and 2 m (6^1/$_2$ ft) in the Arctic. Other factors (discussed below) play a role in altering the thickness, so that ice floes in excess of 10 m (33 ft) are encountered in both the Antarctic and Arctic. However, these masses are a result of deformation processes rather than due to normal undisturbed ice growth (p. 28).

What happens to all that salt?

The salts within seawater cannot be incorporated into ice crystals – when ice crystals grow the salts are expelled. Therefore, in the growing skeletal ice layer the blades of ice are separated by grooves filled with brine expelled from the ice crystals. The brines may be released into the water below, or trapped into inclusions in the ice as the ice consolidates. The input of cold salty water into underlying waters gives rise to cold, dense water masses that sink to the bottom of the ocean and are fundamental in driving ocean circulation patterns (p. 42).

A complex interaction between temperature, salinity, ice-crystal growth and the distribution of the concentrated brine gives the resulting block of ice a specific microstructure. In a rather crude way, sea ice can be considered to be rather like a sponge or Swiss cheese – an intricate solid matrix permeated by a labyrinth of channels and pores that contain the highly concentrated brine. The brine channels vary in size from a few micrometres through to several millimetres in diameter, and they connect the brine inclusions that were created when the ice was forming.

The amount of ice in relation to the channel volume increases proportionally with decreasing temperature. This means that more salts are concentrated into less space. As a consequence, the salinity of the remaining brine will also increase with decreasing temperature. For example, for ice made from water with a salinity of 34, the corresponding salinity of the brine in the ice will be 100 at –6°C (21°F) and 145 at –10°C (14°F).

A block of cold ice will have a smaller volume of brine channels than the same block of ice warmed up. The brine channels of the cold ice will be fewer in number, smaller and have fewer interconnections than in the warmer ice. Ice at the top of an ice floe is usually colder than ice in contact with the seawater. There is, therefore, a temperature gradient throughout the floe, and this imparts a gradient in the salinity of the brine and the overall volume of brine within the ice.

As the salinity increases and the temperature decreases, several minerals become oversaturated at different temperatures and precipitate out of solution. So the downward shift in temperature not only alters the concentration but also the composition of the brine within sea ice, which clearly influences biological activity taking place within the ice matrix.

As ice gets older several processes gradually result in its further desalination. Drainage of dense, cold, concentrated brine simply by gravity is the main mechanism by which this happens, although expulsion of brine from the pores under pressure through cracks and fissures is also thought to occur. Another way in which ice is desalinated occurs within ice that has melted snow or ice water on the surface. The low salinity meltwater percolates down through the ice floe, displacing the brine and flushing it downwards. This meltwater flushing is an important mechanism by which ice is desalinated, especially in thick Arctic ice

Below
Resin cast of the labyrinth of brine channels that permeate sea ice.

where extensive surface melting takes place in ice that lasts more than one season. Therefore ice, especially sea ice that lasts several seasons, can in fact be a source of freshwater, since it has been desalinated with time. Naturally, since sea ice is generally cooled from above, the ice towards the top of the floe is freshest – as long as it hasn't been flooded by seawater.

Flowers on top of the ice

Not all brine losses from the ice are in a downwards direction. Frost flowers are fragile ice crystals covered in salt that grow to a height of 1–3 cm ($^1/_2$–$1^1/_4$ in) on snow-free surfaces of young sea ice in both the Arctic and Antarctic. The source of the flowers is brine expelled from the very upper surface of the ice, and evaporation of vapour from the brine gives rise to conditions in which these beautiful structures can form. The flowers are very salty due to expelled salts, and melt quickly because of their high salt content. They can therefore soon turn to a very salty slush when warmed by the Sun. Salt flowers can form in such densities that they actually form an insulating layer, and the temperature of the ice under the flowers can be one or two degrees warmer than adjacent bare ice.

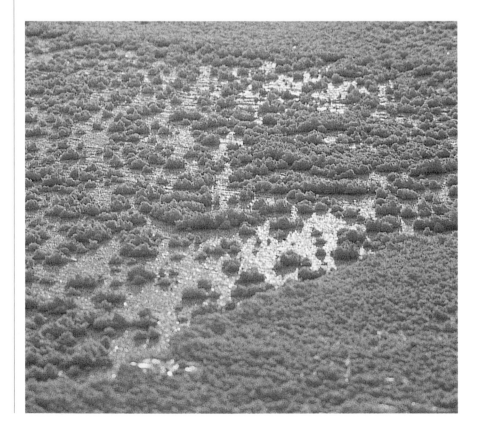

Right
As brine is expelled from a growing ice sheet, frost flowers can develop on the ice surface.

Bubbles in the ice

Seawater is saturated with atmospheric gases such as molecular oxygen, carbon dioxide and nitrogen. The amount of these gases that can be dissolved in seawater depends on the temperature of the water. In cold water more gas can be dissolved than in warmer water. However, the solubility of gases in seawater is not just a function of temperature, it also depends on the salinity of the water (reducing as the salinity increases). Naturally the dissolved gases also become incorporated into a growing ice sheet, and are subject to the same processes as other dissolved constituents such as salt. As the ice grows there is generally a degassing of the ice so that elevated concentrations of oxygen, carbon dioxide and other gases can be measured in the waters immediately below a growing ice sheet.

Above
Bubbles of gas can collect within sea ice.

As the brines associated with sea ice become more saline with decreasing temperature, the combined temperature and salinity effect on gas solubility leads to degassing of the brines as the dissolved gases are forced out of solution. It has been suggested that in cold ice so much gas may be forced out of the ice that oxygen levels become critically low to the point that there is none (anoxia). It is a curious thought that on the surface of the oceans we may in fact have large areas where there is little or no oxygen. However, biological activity within sea ice may dramatically override such physical changes, and in many studies the ice is in fact a place where high oxygen concentrations accumulate (see p. 96).

Pack ice is not just a flat sheet

Sea ice is often described as being like a skin on the surface of the ocean. But unlike a skin it breaks apart and cracks due to the stresses and strains of water movements, tides and wind. However, it is not fragile like glass and even a small ice floe a few metres in diameter and less than 15 cm (6 in) thick is more than enough to support a seal, weighing the best part of a tonne.

There is a rather interesting example of how polar bears venturing on to young ice less than 10 cm (4 in) thick seem to be aware of the physical properties related to the strength of sea ice. On the ice, the bear sets down each paw just long enough for the maximum stress to be reached (a few seconds at the most), and then lifts it again to place it ahead on unstrained ice. This allows the bear to cross regions of ice that would never support the weight of the bear if stationary for more than a few seconds.

Pack-ice fields are dynamic

Pack ice is a dynamic environment. Surface currents and winds transport the ice over thousands of kilometres, although when you are in the middle of a field of pack ice it is sometimes difficult to imagine that you are not in a static and solid landscape. On the other hand, it is an eerie feeling to stand on an ice floe and feel it rocking gently due to the waves underneath. It certainly reminds you that there is an ocean beneath your feet that may be 6 km (4 miles) deep.

In fact ice floes are constantly being jostled around and although flat expanses of nilas ice may be a feature of young ice fields, in a region of ice that has been around for a few months, the landscape is frequently more chaotic. There may be floes rafted on top of each other, rows of ice boulders, all interspersed with areas of open water.

Looking for leads

Wind is a very effective force for breaking open fields of pack ice. As soon as the ice breaks apart, the water underneath is exposed and this is known as a lead. Naturally the ice gives way at the weakest parts of the pack, either through thin ice floes or between ice floes. Depending on the strength of the wind these leads can cover areas of just a few square metres, before being closed over again by moving ice floes, or they can extend for many hundreds of kilometres, effectively producing seas surrounded by a coastline of ice. Seafarers often negotiate their way through pack ice by carefully following such leads. When a vessel becomes stuck in the ice, sometimes the only thing to do is wait for favourable shifts in the wind that will open up a lead to allow escape. This can take hours, days or even months as was clear from the well-publicised incarceration of the Russian Icebreaker *Magdalena Oldendorff* in the Antarctic pack ice from June to November 2002, although many of the crew were airlifted off the ship in several rescue attempts made by a variety of other icebreakers.

When a lead opens up there is a tremendous temperature difference between the air and the water. The freezing air, down to $-40°C$ ($-40°F$), is much colder than the water, which at its coldest is $-1.86°C$ ($28.65°F$). The evaporation and condensation of the surface waters gives rise to frost smoke so that a recently opened lead in winter looks as though it is steaming or smoking.

Of course, the leads will freeze over, and young ice can rapidly cover a lead. The same processes described for new ice in winter will take place within the lead. Therefore in calm conditions nilas ice will form, but if there is considerable wind action pancake ice will form. For these reasons alone there will be a range of ice types within any region of pack ice.

Polynyas

Sometimes large bodies of water within the pack ice persist longer than leads. These polynyas (derived from the Russian word for open or hollow) can persist throughout the whole of the winter, and may occur in the same region over a number of years. Sometimes recurring polynyas open up at the same time each spring, and are so predictable that they are important features for the seasonal activities of Inuit hunters.

Polynyas can be formed either by persistent winds and/or ocean currents sweeping away new ice formed in an area of ocean. These 'latent heat' polynyas often form along coastlines or are associated with islands and even large grounded icebergs. Polynyas can also result from warm subsurface water rising up to increase the temperature of the surface water above the freezing point. Such 'sensible heat' polynyas are often linked to changes in the underwater topography of the seafloor causing the warmer waters to rise.

Above
Sea smoke occurs when very cold air passes over relatively warm water.

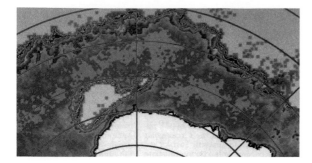

Polynyas vary greatly in size from a few square kilometres to huge areas with large implications for the oceanography and wildlife living there. In the Arctic, the recurring North Water Polynya covers an area up to 50,000 km² (19,305 sq miles). One of the most well known Antarctic polynyas was the Weddell Sea Polynya that opened up in the Weddell Sea each year from 1974 to 1976, covering an area of 350,000 km² (135,135 sq miles).

Above
The massive area of open water, the Weddell Sea Polynya, was a common feature of the pack ice in the 1970s.

Opposite
When ice floes are pushed by water currents or wind the ice can be deformed to form extensive pressure ridges.

Pressure ridges

As well as pulling ice floes apart, wind and ocean currents can also cause them to collide. As they collide the converging edges of the ice masses break into boulders of ice and the rubble collects in ridges of ice called pressure ridges. Tremendous volumes of ice, up to thousands of tonnes, can collide along transects of tens to hundreds of kilometres. These jumbled up blocks of ice can extend down (keel) to 50 m (164 ft) under the water and tens of metres into the air (sail), although these dimensions are extreme. Such substantial sails and keels clearly are important in determining the effects of currents and wind on the subsequent transport of that ice field.

So both convergent (pressure-ridge formation) and divergent (lead formation) processes result in pack ice having a non-uniform thickness. In fact, because 30–80% of the volume of an ice field may be contained within pressure ridges, the dynamics of that ice field caused by wind and ocean currents can be more important to the amount of ice produced than the simple freezing of water by thermodynamic processes.

Right
Schematic illustration to show how convergent conditions result in pressure-ridge formation, and divergent processes result in areas of open water, with the possibility of new ice forming in opened regions.

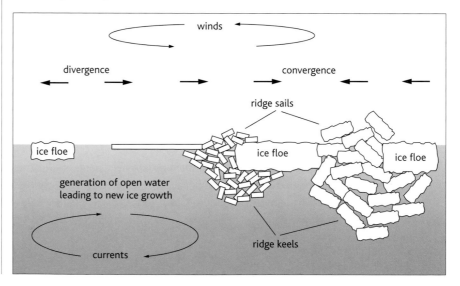

Porridge ice

When there is a lot of snow and the ice is rather thin and wet, such as in the case of new ice, the effects of these dynamic processes can be quite different. Instead of creating impressive solid structures, the snow, ice and water mix together to form a stiff mixture of slush ice that has the consistency of thick porridge. This porridge ice is particularly difficult for ships to manoeuvre through. Porridge ice tends to envelope a ship in a viscous layer that simply will not give way. As a consequence it can stop even some of the most powerful ice-breaking vessels capable of breaking through several metres of hard ice.

Sea ice doesn't last forever

Put an ice cube into a glass of water, or in the sun, and it melts. The same is true of sea ice in the oceans. Sea ice is mostly an ephemeral feature of polar oceans and seas, and its formation, consolidation and subsequent melt is one of the major features that give these systems their highly seasonal dynamics. As discussed below, this seasonality has massive implications for physical processes as well as the biology of the creatures that inhabit these regions.

Opposite top
Sea ice is only an ephemeral structure eventually melting into the surrounding sea.

Below
Sticky porridge ice can result in ships making little progress through pack ice, even though the ice is not thick.

Ice edges

The outer extent of the pack-ice zone is defined by the latitude at which the sea ice comes into contact with warmer waters and/or warmer air temperatures. Therefore, it is a consequence of water movements, day length and total solar radiation (a function of day length, the angle of the sun and cloudiness). The actual ice edge is also strongly influenced by prevailing winds. But there is rarely a transition from 100% sea-ice cover to open water. Instead there is a gradual progression from closed pack ice through to open water. Within this transition is a zone referred to as the marginal ice zone (MIZ), a region of broken ice floes and open water that can extend for 10–100 km² (4–40 sq miles). The concentration of sea ice is often described on a scale of 0 to 10: open water is 0 and 100% ice cover with no open water between the floes is 10.

Below
When temperatures increase, the closed pack ice rapidly breaks up into large individual floes.

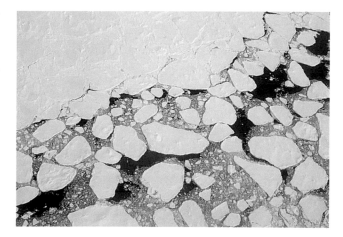

Albedo

Solar radiation is, of course, a major parameter governing how fast a portion of ice will melt. A key factor is the amount of energy actually absorbed by the ice. Albedo is the ratio of reflected to incident radiation and is therefore a measure of the fraction of irradiance that is reflected from the surface of an ice floe. It is a highly pertinent consideration when rates of ice melt are being calculated. Open water has a low albedo (0.05) compared to white new snow, which has a high albedo (0.9).

High albedo is critical to maintaining the integrity of ice sheets and sea-ice cover. For instance, the massive ice sheet overlying the continent of east Antarctica actually receives the maximum monthly input of solar energy of anywhere on Earth. However, since most of this is reflected back (it has a high albedo), it still remains one of the coldest places on the planet.

Ice with snow cover will reflect light more effectively (has a higher albedo) than ice without snow, and therefore melts at a slower rate. Even factors such as whether the snow is wet or dry play an important role, dry snow having a greater albedo than wet snow.

However, many properties of the ice itself influence its albedo. Bubbles of gas or brine pockets increase the scattering and reflection of light, thereby increasing albedo and reducing energy absorption and melting. Thin ice has a lower albedo than thicker ice. New dark ice has a much lower albedo, and therefore absorbs energy to a higher degree than older ice, which therefore melts at a slower rate. If there are sediment particles in the ice or significant growth of algae in and on the ice, these will also effectively reduce the albedo of the ice, increasing the rates of melting compared to particle-free ice.

When ice melts, the albedo of the ice is reduced and therefore more energy can be absorbed. This in turn will increase the rate of melting. This is termed a positive albedo feedback mechanism where the absorption of heat energy leads eventually to an even greater absorption of energy. This happens on a seasonal basis during the melting of sea ice, but more dramatically it is thought that such positive feedback may play a large role in changes brought about by global climate change. Potentially if there is an increased melting of sea ice in polar regions, as in the Arctic, the reduced albedo will induce further warming of the surface waters and thinner ice resulting in accelerated ice melt.

Ponds on the ice

When surfaces of ice floes begin to melt, a common feature is the formation of surface melt ponds. These are a more common feature of Arctic sea ice than Antarctic sea ice, although they can reach high numbers even there. The major reason for this difference is that a greater proportion of Antarctic sea ice is melted by heat derived from the underlying water. In the Arctic the melting from above (from solar radiation) is of a greater significance.

Once a melt pond is established on the surface of an ice floe it clearly has a lower albedo than the surrounding ice, so absorbs more energy, and this enhances melting further. In this fashion melt ponds grow in area and depth. Melt ponds do refreeze over when air temperatures are low enough, and at night melt ponds frequently form a surface layer of thin ice, which then melts again the following day. Because these ponds are largely derived from snow and low saline surface ice, they really are akin to lakes floating on the surface of the ocean.

Eventually the melting reaches a stage whereby all the ice underlying the pond is melted away and there is a direct connection between the ocean and the freshwater contained within the melt pond. This frequently happens at a thaw-hole located at the melt pond's deepest point.

Meltwater can either mix with the surrounding/underlying seawater, or form a freshwater layer at the surface of the ocean. In the Arctic these freshwater layers sometimes collect as lenses, or pools, of freshwater under large sea-ice floes. These can become refrozen into the ice floe in the following autumn, forming a distinctive layer in the ice.

Of course, not all the melting takes place by absorption of solar radiation. The ice can melt from below, through the warmth of the underlying seawater. As the ice warms it can take on a rather fragile appearance, melting first in brine channels containing high salinity brine. At the extreme case the ice becomes so porous that a skeleton remains, with a molten ice matrix surrounding a channel system that is filled with seawater. The term rotten ice is quite appropriate, since such ice has very little strength and is easily broken up.

Below
Melting of snow and ice surfaces can result in the formation of melt ponds.

Other melt features

Many melt ponds are initiated by the flooding of ice-floe surfaces by the surrounding seawater. These floods either wash the snow away or moisten the snow so that its albedo is reduced. Flood-induced surface ponds are also often associated with uneven surfaces on the floes that cause floodwater to collect into ponds, and subsequently induce growth of a larger melting event. A common place for this to happen is in parts of ice floes where pressure ridges form very uneven and varied surface terrains.

A common feature of Antarctic sea ice in austral (southern hemisphere) summer is the formation of layers of rotten surface ice between the snow-ice interface or in the top 50 cm (20 in) of the ice floes. These take various forms

Above right
Melt ponds often first form around deformed pressure ridges on the ice.

Right
Flooding by seawater and melting of surface ice layers can result in surface gap layers that are rich in biology.

such as semi-continuous gaps or voids. These can be filled with rotten ice slush, frequently mixed with seawater that percolates into the floe through these gap layers.

These gaps are frequently associated with the flooding and refreezing of snow layers to form snow-ice. Such processes result in differential melting properties between the hard freshwater snow ice and more saline sea ice below. Often when moving through a region of pack ice where there is a lot of this superimposed snow-derived ice, slabs of the snow ice fall from ice-floe surfaces as the ship disturbs them.

Black ice

When ice is formed in coastal regions and shallow waters it is possible for large concentrations of sediments from the sea floor to become incorporated into the new ice. This is especially true of Arctic coasts where in a year large rivers transport thousands of tonnes of particles from the land to the ocean. A whole range of particle types and sizes, from clay to larger soil pieces, can be incorporated into the ice and even tree trunks have been found in Arctic sea ice.

The most obvious effect of sediment in the ice is to discolour the ice. Where lots of sediment is trapped in the ice it can look black rather than white. When this happens the albedo is significantly reduced and the rates of ice melt increased. The concentrations of sediment in Arctic sea ice can range from near zero per litre of ice to over 10 g per litre in very heavily sediment-laden black ice. In one observation gravel and boulders were seen on an ice floe more than 50 km (31 miles) off the east coast of Greenland. Apparently it looked as though the debris had been dropped on the floe by a truck.

Below
Sediments can become incorporated into sea ice as it grows.

Another significance of sediment incorporation into sea ice is that the sediment can be carried many thousands of kilometres as the ice drifts from where it was formed. Normally sediments from rivers fall out of the water to be deposited on the sea floor in close proximity to the coast. However, sediments caught up into the ice can be deposited on the sea floor of much deeper waters.

Sediment-laden ice is a feature of sea ice in many regions of the Arctic Ocean, but it is seldom encountered in the Southern Ocean. This is because there are very few shallow coastal waters where sediments from the sea floor can be suspended into new ice and, of course, there are no rivers of significance on the Antarctic continent to introduce sediment particles to the surface waters. The 30 km (19 mile) long river Onyx does not drain into the sea, although the river Alph does sometimes discharge debris into McMurdo Sound. Black icebergs are seen in the Antarctic, however. These bergs have high concentrations of sediments that were incorporated as the glacial ice scoured rock surfaces before calving into the sea.

It's all a question of energy balance

Ultimately the rate at which sea ice grows, the time it remains consolidated, and the rate at which it subsequently melts is a function of the energy balance in the region. Solar energy is a major heat source and the amount of solar energy reaching the ice depends on day-length, the angle of the sun and the amount of clouds present, since clouds are superb barriers to incoming radiation. Since the Earth is closer to the Sun in the Antarctic (austral) summer, but further away in the Arctic (boreal) summer, more energy enters the Antarctic. In addition, the Arctic atmosphere contains more radiation-absorbing particles, dust and a greater cloud cover. All these factors add up to the Antarctic receiving about 15% more solar energy than the Arctic over the year.

As mentioned, albedo is central to the heat balance of a region. Generally albedo values are greater in the Antarctic compared to the Arctic, largely due to the differences in snow characteristics in the latter, including the wetness and thickness of the snow.

Greenhouse gases also have an impact on the energy balance of a region. When an object is warmed by solar energy, thermal energy is then lost by the emission of infra-red radiation. Clouds, atmospheric water vapour, nitrous oxide, methane, carbon dioxide and chlorofluorocarbons (CFCs), the so-called greenhouse gases, absorb this radiation. This trapping of energy is called the greenhouse effect, and without it the Earth would be 30–40°C (50–70°F) cooler than it is. Therefore, in itself, this effect is vital to life on Earth as we know it. Water vapour is the greatest contributor to the effect (about 60%), but carbon dioxide can account for up to 25%.

Over the past two centuries man's activities have increased the loading of carbon dioxide and other greenhouse gases into the atmosphere, to the extent where it is thought to have caused much of the global warming that is now well documented and of concern to us all. There is evidence from long-term data sets that global warming over the past three decades is having an effect on sea-ice distributions.

The atmospheric absorbers warm up and in turn emit long-wave radiation, a proportion of which will be returned to Earth and reabsorbed. When the energy emitted from an object is greater than that received by back radiation from above there is a net energy loss. There is a net loss of energy by radiation in the poles and net gain in equatorial latitudes, and in general the net energy loss in the Antarctic is greater than in the Arctic, which is partly due to the greater prevalence of clouds in the latter, coupled to the humidity and chemical composition variations between the two systems.

Therefore the poles can be considered as sinks for thermal energy whereas the equatorial regions are sources. This is a major driving force for climate and ocean circulation patterns, with the heat lost by the cooling of air and water in the polar regions being replaced by warmer air and water from lower latitudes.

First-year and multiyear ice

Not all the ice that is formed in winter in both the Antarctic and Arctic actually melts again the following year. Ice that survives through to a following year is referred to as first-year ice, and ice that survives two summers is referred to as multiyear ice. In much of the Arctic the ice is multiyear ice lasting several years, whereas most of the Antarctic pack ice is no more than a few months old.

The longevity of sea ice depends on the wind and ocean currents that determine its passage through the ocean. In the Southern Ocean, the ice fields are generally being pushed away from the continent into lower latitudes and therefore warmer conditions. In general 80% of the ice formed each year melts again within six months. Only in the Weddell and Amundsen Seas do significant amounts of ice last for more than one year.

Because the Arctic Ocean is surrounded by landmasses, and there is little contact with other oceanic systems, much of the sea ice lasts more than one season (up to six years). The sea ice is transported in gyre systems (great clockwise circular motions of water) or in the trans-Arctic drift (p. 51) and out of the Fram Strait where it meets the North Atlantic and finally melts.

How thick does pack ice actually get?

Several different factors have been mentioned above that influence the thickness of sea ice. Although large-scale differences between the Arctic and Antarctic will be discussed elsewhere, it is pertinent to relate differences in sea-

ice thickness between the polar oceans to the main factors governing the ultimate thickness that ice obtains.

Overall, Arctic ice is greater than 2 m (6½ ft) thick. In some regions such as along the coasts of northern Greenland and the Canadian archipelago, it is estimated that the average thickness is 6–8 m (20–26 ft). The ice of the Southern Ocean is considerably thinner at less than 1 m (3 ft). Thick, heavily ridged ice floes greater than 10 m (33 ft) are found in Antarctic waters. But these are relatively few and insignificant when considering the overall distribution of Antarctic sea ice.

The main difference between the two regions is that the surface layers of the Arctic Ocean are colder than those of the Antarctic. There is a great deal of river runoff into the Arctic (none in the Antarctic) and this low salinity surface layer forms an effective barrier between the warmer water layers below. The surface of the Southern Ocean is not cut off to the same order of magnitude, and there is a considerable flux of heat into surface waters from below.

As mentioned above, the snow cover on Arctic ice tends to be considerably thinner than that on Antarctic sea ice. Although this results in a lower albedo with increased surface melting of Arctic sea ice, the insulating effects of the snow means that Antarctic sea ice generally grows at a slower rate than Arctic sea ice.

Above
The Arctic Ocean is surrounded by landmasses that govern Arctic pack-ice dynamics.

Older sea ice in the Arctic

As mentioned, there is a higher proportion of multiyear ice in the Arctic compared to the Antarctic. The older the ice, the more chance it has to thicken due to pressure-ridge and rafting events. Deformation processes such as pressure-ridge formation are much greater in the Arctic Ocean than in the Southern Ocean. In the latter, ice drift is mainly divergent, the ice drifting towards the open ocean. This means that leads are more likely to form in the pack ice of the Southern Ocean than in the Arctic and so there is a higher proportion of thinner, new ice contributing to the total volume of ice.

Another major difference between the Arctic and Southern Oceans is that most of the Arctic ice forms at latitudes north of 70°N, whereas in the Southern Ocean a large percentage of the sea ice extends northwards to latitudes of 60°S. The consequences of this are that for much of the Arctic pack ice, air temperatures are lower, incoming solar radiation is less and the length of the summer season is shorter than in the Antarctic sea-ice zone.

Opposite (top and bottom)
Much of the Arctic ice is heavily deformed ice several metres thick.

Below
A lead in the Antarctic pack ice freezing over.

Frozen water and deep ocean circulation

The expulsion of large amounts of cold brine from sea ice in the Labrador and Norwegian-Greenland seas in the northern hemisphere, and in the Weddell and Ross Seas in the southern hemisphere, results in the water beneath the ice being cooled and laden with extra salt. This increases the density of the water and it sinks rapidly to great depths. The production of high density seawater is called deep-water formation, or bottom-water formation, and is the basis of a process that drives the deep circulation throughout the world's oceans. This is very different from the surface ocean currents and circulation patterns that are largely driven by winds.

Despite their very different characteristics and the vast expanses covered by the oceans of the world, they are all interconnected by a large-scale movement of water that is referred to as the Global Thermohaline Conveyor Belt. The basis of thermohaline circulation is that a kilogram of water that sinks from the surface into a deeper part of the ocean, displaces a kilogram of water from the deeper waters.

In the conveyor belt circulation in northern latitudes, warm surface and intermediate waters – to depths of 1 km ($^2/_3$ mile) – are transported towards the North Atlantic, where they are cooled and sink to form North Atlantic deep water that then flows southwards. In southern latitudes rapid freezing of seawater also produces cold high-density water that sinks down the continental slope of Antarctica as Antarctic bottom water. These deep-water masses move into the southern Indian and Pacific Oceans where they rise towards the surface.

Right
The sinking of cold saline water in the Arctic and Antarctic Weddell and Ross Seas, drives large-scale ocean circulation in the so-called 'Global Thermohaline Conveyor Belt'. (Dark blue, cool ocean current; light blue, warm ocean current).

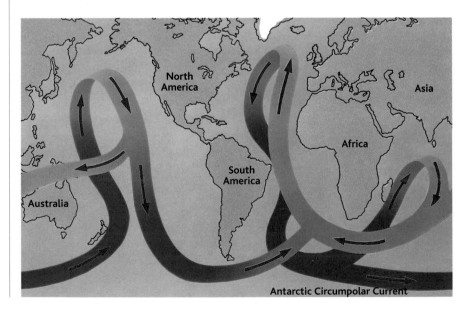

The return leg of the conveyor belt begins with surface waters from the north-eastern Pacific Ocean flowing into the Indian Ocean and then into the Atlantic.

Transport of oxygen, carbon dioxide and nutrients

The deep waters of the polar regions are oxygen rich, and so are fundamental in transporting oxygen to the ocean depths where it is used by deep-sea organisms. The sinking of the cold, well-oxygenated saline water is often referred to as ventilation of the ocean's depths. Dissolved organic matter and inorganic nutrients are also transported in this way, with bacteria releasing inorganic nutrients from organic matter en route. Therefore water rising at the end of the conveyor belt in the north-eastern Pacific has a higher nutrient loading, and lower oxygen concentration, than North Atlantic water at the beginning of the conveyor belt.

The sinking of surface water in the far North Atlantic and in the Weddell and Ross Seas continually renews the water exposed to the surface. Cold, carbon-dioxide poor water at the surface tends to take up carbon dioxide from the atmosphere, but sinks before equilibrium between the concentration of carbon dioxide in both water and air is reached. This is one of the reasons why the polar regions are considered to be sinks for carbon dioxide: carbon dioxide will be absorbed into the under-saturated surface waters in polar oceans, whereas carbon dioxide is released into the atmosphere from the super-saturated surface waters in equatorial regions.

The explanation of the conveyor belt mechanism is clearly a gross oversimplification of all the complicated patterns involved in the circulation of the world's oceans. However, it does serve to describe the major driving forces that can trap water masses in the deep sea for up to 5000 years, or possibly longer, before allowing them to re-surface.

Pack-ice regions of the world

3

THE REMOTENESS AND HOSTILE NATURE OF POLAR REGIONS
HAVE ATTRACTED EXPLORERS FOR THOUSANDS OF YEARS.
HOWEVER, IT IS EXACTLY THESE CHARACTERISTICS THAT HAVE
LIMITED OUR UNDERSTANDING OF THE MAJOR PROCESSES
GOVERNING THE DISTRIBUTION AND PHYSICAL ATTRIBUTES OF
THESE OCEANS. EVEN A COMBINATION OF AIRCRAFT AND SHIP
SURVEYS ACHIEVES A VERY LIMITED SPATIAL RESOLUTION.

T HE RESULTING INFORMATION is on a few thousand square kilometres at the very best. In order to obtain a bigger picture of seasonal ice dynamics scientists have turned to viewing the ice from space. Using the latest satellite technologies it is possible to monitor daily changes in sea-ice cover over the whole of the Arctic and Southern Oceans with such accuracy that differences of less than 25 km (16 miles) can be seen.

At the frontier of this research are satellites that are able to measure the microwave radiation emitted by land, ice and water. This is a passive measurement, in that the satellite systems do not emit any power to the target; they simply measure the energy emitted by an object. The results show a stark contrast between ice and water, making ice distribution maps relatively easy to compile. Microwave frequencies of radiation can pass through clouds and the rest of the atmosphere, and are not dependent on daylight. Therefore, microwave data can be collected day and night and in just about any weather.

There are also satellites that emit their own electromagnetic radiation, sending a series of signals to the target and then detecting the echoing microwave radiation. By using this active technique it is possible to obtain a higher spatial resolution than using the passive microwave techniques, and there is the advantage that these methods are transparent to cloud and can be used at night.

Even better resolution can be achieved using instruments borne on satellites that can measure visible and infra-red radiation. For example, data collected from a visible channel of advanced very high resolution radiometers (AVHRR) can routinely achieve a better resolution than 7 km ($4^1/_2$ miles), though these

Opposite
Despite the best endeavours of researchers, the area of ice they investigate is rather limited.

Right
Images of summer minimum and winter maximum sea-ice extents in the Southern Ocean, from AVHRR and SSMI sensors on satellites.

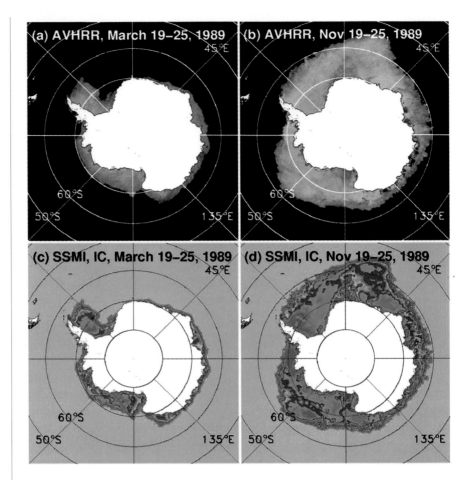

measurements are restricted to daylight hours. Infra-red measurements can be made at night, but both visible and infra-red radiation are blocked by clouds and so can only be measured on clear days.

A long data set has been obtained using passive microwave data collected by a range of sensor types including: special sensor microwave imagers (SSMIs), scanning multichannel microwave radiometers (SMMRs) and advanced microwave scanning radiometers (AMSRs) that have been launched on many satellites since the 1970s. Scientists now have a complete record of passive microwave data showing sea-ice distribution and extent from 1978 to the present day. This has given us a huge advance on our knowledge of sea ice, which had previously relied on the limited information recorded in ships' logs on the relatively few ships that were capable of sailing in pack-ice regions. By default these records were mostly restricted to ice-edge regions, where the ships could operate and/or where the seals, whales and fish were to be found (the main reason for much of the commercial operations in these waters).

How can we measure the thickness of the ice from space?

One of the major problems with satellite observations of sea ice is that even at their best they only provide information about the ice surface. In particular the microwave maps do not even provide the resolution to determine small to medium-sized leads in the ice, and only the largest of ridges are discernable. Up to now we have not been able to gain large-scale information about the thickness of the ice below the surface, which is a pity because it is not just the aerial extent that is of significance. When estimating the impact on large-scale climate and oceanographic processes, it is more useful to know the volume of ice that is formed and subsequently melts.

Satellites can be equipped with laser altimeters that measure the height of the satellite above the Earth. Since the orbital height of the satellite is constant, it is possible to measure the height above the ice and the height above water. Comparing the two gives a measure of how high the ice is floating above the water (its freeboard). This can then be transferred, using proven mathematical calculations, to produce a measurement for the thickness of the sea ice. CryoSAT (due to be launched in 2004), and IceSAT (launched in 2003) satellite missions, by the European Space Agency and NASA, respectively, hope to use the latest altimetry methodologies to observe high resolution ice thickness.

How much sea ice is there?

At any one time sea ice covers an area of 17.5–28.5 million km^2 (6.8–11 million sq miles) which is equivalent to 3–6% of the Earth's surface. In the Arctic, the extent of the sea ice varies from a September minimum of 7.5 million km^2 (3 million sq miles) to a March maximum of 15.5 million km^2 (6 million sq miles). In the Antarctic the minimum occurs in February at 3.8 million km^2 (1.5 million sq miles) increasing to a maximum of 19 million km^2 (7.3 million sq miles) in September.

Below
The satellite CryoSAT due to be launched in 2004 will determine the variations in ice thickness around the world.

Right
The maximum and minimum extent of sea-ice cover in the Arctic Ocean (top) and Southern Ocean (bottom) in 2000.

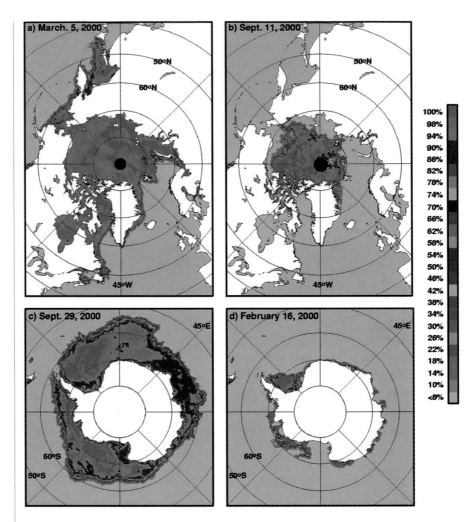

Antarctic and Arctic seasonal changes

In the Arctic the period of sea-ice growth is about equal to the time it takes for the subsequent seasonal decay of ice. In contrast, the ice in the Southern Ocean grows for about seven months and decays over the following five months. The main reason for the difference is that in the Antarctic the melting process is accelerated by warm waters surrounding the whole ice mass, while the Arctic is surrounded by land and has little contact with warming waters.

It is interesting to note that the lowest values in total sea-ice cover on the Earth occur in February (winter in the Arctic and summer in the Antarctic), and the maximum in November (spring in the Antarctic and autumn in the Arctic). The November peak is a result of the longer growth period of Antarctic sea ice compared with its decay. Interestingly between June and November, the total

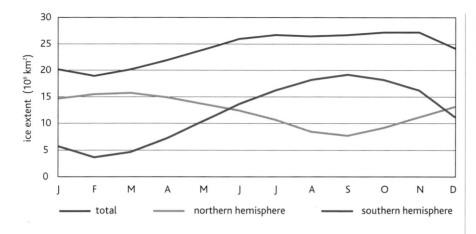

Left
Monthly average variations in sea-ice extent in the northern and southern hemispheres and total sea ice on Earth.

sea-ice cover on the Earth is rather constant, which is staggering considering the vast amounts of ice growth and decay that is going on in the two hemispheres.

Although some differences between the Arctic and Southern Oceans have been discussed briefly already, it is pertinent to give a little more detail about the major ocean currents and key features of the systems that ultimately govern some of the phenomena described above. It is important to keep in mind that sea ice is overlying water masses, some of them with huge spheres of influence way beyond the polar regions. The mixing of ocean currents with polar waters is central to the rate of ice formation and subsequent melt.

The Arctic Ocean

The Arctic Ocean is an almost closed basin, with very little exchange with the surrounding large oceans. The major exchange of water takes place through the deep Fram Strait that lies between Svalbard (often called Spitsbergen in the literature) and Greenland. There are other connections with the Atlantic and Pacific Oceans through the Barents Sea, the Canadian Arctic archipelago and Bering Strait. However, these are far shallower than the Fram Strait and are the site of much reduced exchange. The deepest part of the Arctic Ocean is over 4 km (2^1/$_2$ miles) deep, although over one third of the ocean is less than 100 m (328 ft) deep, and the resulting average depth of 1800 m (5900 ft) makes it the shallowest of the oceans. The continental shelves fringe the surrounding landmasses and in some regions shallow seas no deeper than 50 m (164 ft) can extend up to 600 km (373 miles) from the land.

In the Arctic the major heat input is provided by northward moving warm air that interacts with cold polar air in cyclones at around 60°N. The air then travels northward high on the troposphere subsiding near the pole, returning as surface winds in a generally easterly direction due to the Coriolis force.

Below
The Arctic Ocean showing minimum cover of sea ice.

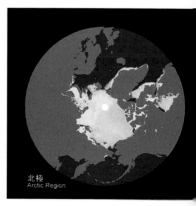

Right
Schematic illustration showing the main currents in the Arctic Ocean that carry sea ice.

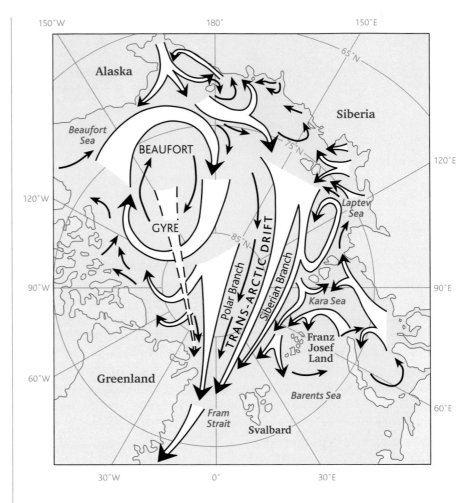

Arctic water masses

Warm waters are carried into the Arctic Ocean predominantly by the west Spitsbergen current, an extension of the Norwegian Atlantic current that passes from the North Atlantic through the Fram Strait following a deep trench into the Arctic Ocean. Some 60–250 million km³ (14,000–60,000 million cubic miles) of water per year are transported by this route. A much smaller amount, of around 25,000 km³ (6000 cubic miles) per year, enter by the Bering Strait. These relatively warm, 3°C (37°F), and saline waters (salinity of around 35) tend to sink underneath the colder polar surface waters which have a lowered salinity (around 30–32) because of the large freshwater input by rivers. The surface waters can extend down to 200 m (656 ft) in some areas such as the Beaufort Sea, although in general they form a layer of low salinity water that is 30–60 m (98–196 ft) deep. This effective barrier prevents mixing with the deeper waters, and therefore heat exchange with the deep waters. In summer the input of vast volumes of warm freshwater from the rivers is, of course, another major source of heat into the Arctic Ocean.

The Beaufort Gyre and trans-Arctic drift

Within the Arctic basin the surface currents, which are mostly driven by anticyclone winds, set up some characteristic features that determine the passage of ice within the ocean. These are the main controlling factors governing the transport of surface waters and, of course, the sea ice. The Beaufort Gyre is an anticyclonic gyre in the Canada Basin that takes between seven and 10 years to make a complete circuit. Therefore there is a lot of ice locked up in this gyre, a key factor in determining the occurrence of large amounts of multiyear ice in the Arctic. The Beaufort Gyre is not always strongly developed. For example, it was well developed in the 1980s but considerably weakened in the 1990s. A weaker gyre clearly results in a greater dispersal of sea ice and less opportunity for thick multiyear ice to develop.

The other main current system in the Arctic is the trans-Arctic drift stream. This transports water and ice from the Eurasian ice shelves, across the pole and eventually links up with the east Greenland current at the Fram Strait. This is the main current out of the Arctic Ocean transporting about 91,000–910,000 km³ (21,832–218,320 cubic miles) of water, and an estimated 4 million megatonnes of sea ice per year. The vast majority of the ice is transported through this passage.

The Fram Strait has been described by some as being the "Arctic's great two-way water highway". The main input of water takes place up the eastern edge of the strait, and the main output is along the western edge. Fridtjof Nansen was the first to test the idea of the trans-Arctic drift, when in 1893 he froze his ship the *Fram* into newly-formed ice off the New Siberian Islands. The ship drifted, missing the North Pole, to eventually emerge in the Fram Strait.

Long-term trends in the extent of Arctic sea ice

Satellite data have shown that the perennial sea-ice cover (i.e. ice lasting through the summer) in the Arctic has been declining at a rate of 9% per year since 1978. If this rate is continued it is estimated that there will be no sea-ice cover lasting through the summer by

Below
The minimum summer ice extent in the Arctic is reducing. Top left, average from 1979 to 1989; top right, average from 1980 to 2000; bottom left, difference between the 1979 to 1989 and 1980 to 2000 averages; bottom right, predicted summer minimum ice extent for the 2050s.

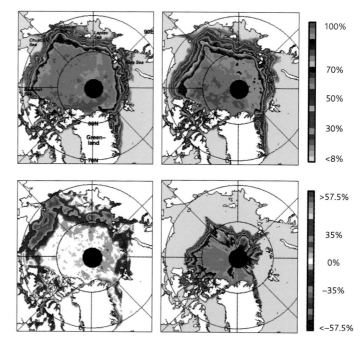

the end of the century. Correspondingly temperatures in summer are increasing by around 1.2°C (2.2°F) per decade. The ramifications are that spring melt will occur earlier and freezing take place later in the year causing further thinning and retreat of perennial ice.

These trends are not uniform over the whole of the Arctic. The greatest amount of ice-melt in summer takes place in the Beaufort, Chukchi, Laptev and Kara Seas, whereas there are accumulations of summer sea ice close to Greenland.

More generally the trends in the extent of Arctic ice averaged over the past three decades are less dramatic than the summer ice trends. There seems to be a more gradual reduction in the ice extent in the northern hemisphere of about 2% per year, although in the central Arctic Ocean this is lower at around 1.2% per decade. Again the trends are not uniform across the Arctic and the Bering Sea shows a 12% increase in sea ice per decade, which is mirrored by a 12% decrease in sea ice in the Okhotsk Sea over the same period.

Arctic thinning

There have been a number of studies demonstrating that the most spectacular changes in the dynamics of Arctic sea ice are not so much to do with the extent, but are concerned with the rapid thinning of sea ice. Comparison of sonar data collected by submarines traversing the Arctic under the ice has been one of the techniques used. By comparing data gathered between 1958 and 1979 with data collected between 1993 and 1997, researchers have been able to detect a 42% decrease in summer ice thickness, especially in the Siberian Arctic and Nansen Basin. There have been other comparisons made of ice-thickness data collected by other submarine campaigns and upward-looking sonars attached to underwater moorings over long periods of time (p. 198). These all confirm that the average sea-ice thickness in the 1990s is significantly less than previously measured.

Estimates of sea-ice thickness trends using satellite-based altimeter measurements of ice freeboard and the estimated ice thickness, have shown convincingly that there has been a thinning of average Arctic ice thickness between 1993 and 2001. The researchers have shown that the sea-ice mass can decrease by a massive 16% within one year (1998–1999). They link these changes to the length of summer melting period, and provide data to show that extending this by just one day results in an extra 4.9 cm (2 in) of ice melt to the average ice thickness. This may sound somewhat insignificant, but multiply that by the millions of square kilometres of ice, and it becomes a highly significant amount of water.

Arctic and North Atlantic Oscillations

Over the past few decades it has become clear that many oceanographic trends in the northern hemisphere are closely linked to the North Atlantic Oscillation

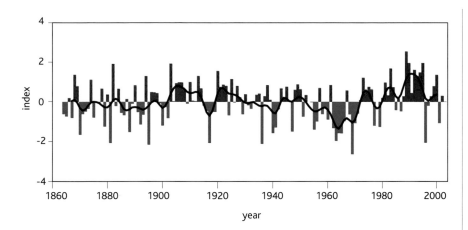

Left
NAO indices have been estimated since the late 19th century. During a positive NAO index there is a strong Icelandic low pressure system and Azores high pressure system; during negative phases the pressure gradient is weak with an Icelandic high and Azores low. NAO index for December to March 1864–2003.

(NAO). On a global scale this is one of the most dominant modes of climate variability following El Niño, although there is very little connection between the two. The NAO compares the atmospheric pressure in the region of Greenland-Iceland with that of the subtropical central North Atlantic in the Azores. The NAO index is defined as the difference between the Icelandic low and the Azores high in winter (December to March). A positive NAO index is characterised by a strong Icelandic low and Azores high pressure, with a corresponding strong north-south pressure gradient. During negative phases the pressure gradient is weak with an Icelandic high and Azores low and a south-north pressure gradient.

During a positive NAO the pressure differences result in stronger and more frequent storms crossing the Atlantic Ocean in a northerly direction. This results in warmer and wetter winters in Europe and the eastern USA and cold dry winters in northern Canada and Greenland.

During a negative NAO the reduced pressure gradient results in fewer and weaker winter storms crossing the Atlantic in a more or less west-east trajectory. They bring moist air into the Mediterranean and cold air to northern Europe and the east coast of the USA, bringing about cold snowy weather. Greenland will have milder winters during these phases.

The Arctic Oscillation (AO) is thought to be closely linked to the NAO. In fact, some researchers say that the NAO is rather a component of the larger-scale AO, and the two terms are often interchanged. From the 1950s until 1979 a negative phase dominated, after which a more positive phase has predominated. There are years when these general trends are reversed such as in 1995 to 1996 when there was a very abrupt reversal of the index.

There has been general warming of surfaces in the Arctic over the past 100 years. However, in the past 20 years the temperature has been increasing eight times more rapidly than the 100-year trend. Many researchers link this rapid acceleration in the warming process to global warming associated with increased concentrations of greenhouse gases. It has also been noted that these rapid warming trends are associated with increasing periods in the positive phase of the AO/NAO. Clearly such AO/NAO trends in wind, storm and warming events will have great influence on the variations in sea-ice distribution in the Arctic basin. For example, when the NAO index shifted from positive to negative during the winter of 1995 to 1996 the sea ice exported through the Fram Strait was reduced by around half. Mathematical models of ice distribution show 28% decreases in ice volume in the eastern Arctic and increases of 16% in the western Arctic when comparing data from 1979–1988 (generally low NAO index) with 1989–1996 (generally high NAO index) data. The Beaufort Gyre was also much weaker in 1989–1996 than in 1979–1988.

The Southern Ocean

The Southern Ocean surrounds the Antarctic continent and is bordered by the Atlantic, Pacific and Indian Oceans. In the Antarctic the continental shelves are generally very narrow, and so the Antarctic pack-ice zone is largely overlying deep oceanic basins 4–6.5 km (2$\frac{1}{2}$–4 miles) deep. Only in the Weddell and Ross Seas are extensive continental shelves present.

The Southern Ocean is bordered by the surface, eastward-flowing Antarctic circumpolar current that transports about 4 million km³ (1 million cubic miles) of water per year. This largely wind-driven current carries the greatest amount of water transported by any of the currents on Earth. Closer to the continent there is a westward-flowing Antarctic coastal current, which largely determines the drift of icebergs that calve from the ice shelves, at least at the beginning of their journeys.

Warm waters from the northern oceans are transported towards the Antarctic continent as circumpolar deep waters that rise to the surface at the Antarctic divergence, close to the Antarctic continent at around 70°S. These waters then move either in a northerly, or in some parts of the Southern Ocean, a southerly direction. The waters moving northwards become less saline and colder as they mix with meltwater from melting icebergs and sea ice. They become low-salinity surface water that continues to move northwards overlaying the circumpolar deep waters. The differences in salinity between the surface and underlying waters are not as stark as those in the Arctic, and although the surface layer may extend downwards for 200 m (656 ft) or so, there is a greater exchange of heat between the two layers compared with the Arctic system.

Moving northwards the maximum extent of this surface water is the Polar Front, which is a boundary where the polar waters meet the warmer sub-Antarctic waters. The Polar Front (Antarctic Convergence) extends around the whole of the continent in a rather constant location, lying within the Antarctic

Below
At its maximum extent the Antarctic continent is surrounded by 20 million km² (8 million sq miles) of sea ice.

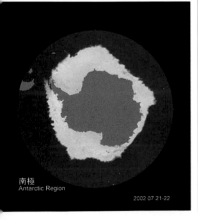

南極
Antarctic Region

2002 07 21-22

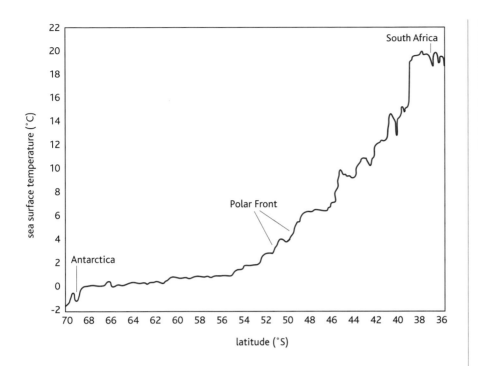

Left
Plot of sea surface temperature travelling from the German Antarctic Neumayer station to Cape Town, South Africa. The steep 5°C (9°F) rise in water temperature between 52° S and 49° S indicates the position of the Polar Front.

circumpolar current. The maximum extent of sea ice does not reach the Polar Front region. The crossing of the Polar Front is always very noticeable, especially when measuring the temperature of the air and seawater. Within a just few hours of a ship steaming towards the south, temperatures can plummet dramatically.

Upwelling circumpolar deep waters that move in a southerly direction become colder on mixing with water from ice shelves and growing sea-ice fields that expel so much salt in brine that the salinity of the water increases. This cold, saline water then sinks rapidly to form Antarctic bottom water spreading northwards along the sea floor. The major sites for this bottom water production in the Antarctic are the Weddell Sea and Ross Sea. Antarctic bottom water produced in the Weddell Sea has been detected as far as 40°N in the Atlantic.

Gyres in the Weddell and Ross Seas

Most of the sea ice in the Antarctic forms and melts again within a single year. However, in the Weddell and Ross Seas there are gyre circulation patterns that result in a certain amount of ice being locked up by surface currents for longer periods of time, and the chance for significant volumes of multiyear ice to establish. These gyres are not complete gyres, as is the Beaufort Gyre in the Arctic, and ice therefore does not circulate round and round the gyre for several years.

The gyres move in a clockwise motion, as does the Beaufort Gyre. However, due to the Coriolis force the ice carried in the surface waters is concentrated towards the centre of the Beaufort Gyre. In contrast, with the Southern Ocean gyres the Coriolis force results in the ice moving outwards, diverging from the centre of the gyre. Therefore, in the Weddell Sea, instead of collecting in the middle, ice is moved outwards in a westerly drift. This results in much of the ice reaching the physical barrier of the Antarctic Peninsula in the west, where ice can accumulate. This causes much of the ice in this region to deform and become heavily rafted, and is a region where some of the thickest sea ice in the Southern Ocean is found.

The trajectory of these gyres can be followed using floats or buoys that transmit signals to satellites. These are either released into water, placed on ice floes or on to floating icebergs. For instance, an iceberg entering the Weddell Sea from the east in the westward-flowing Antarctic coastal current, gets caught up in the Weddell Gyre trajectory, emerging in the west Weddell Sea in the Antarctic Peninsula/Drake Passage region two years later.

Below
Special buoys can be deployed on ice floes. The buoys transmit positional information to satellites enabling their drift to be tracked.

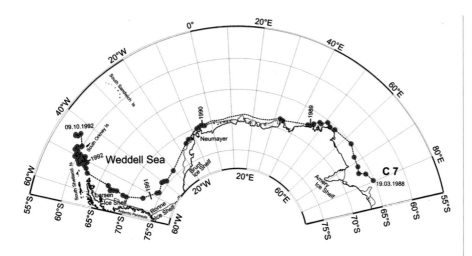

Left
The four-year trajectory of iceberg C7 that was tracked from March 1988 (right of diagram) to its final position in October 1992 (left of diagram).

Antarctic circumpolar ice edge

Of course, the major difference between the pack ice of the Southern Ocean compared with that of the Arctic is that the periphery of the whole ice zone is completely surrounded by water. There will be a complete marginal ice zone (MIZ) permanently present, where melting processes are enhanced by the ocean swell breaking up the floes. The MIZ can extend up to 200 km (124 miles) into the pack. When the ice at the edge melts, huge volumes of freshwater are released into the surface waters that can stabilise the surface waters into shallow, low salinity layers. There are dramatic changes in air and surface water characteristics in the ice-edge regions. It is very noticeable when moving from the open water into the ice that the ice effectively dampens waves and water movements.

Ice edges can be divided into three phases: a growth phase as ice advances and a decay phase where the ice-edge retreats. There is an equilibrium phase where the edge is oscillating between advance and retreat. Ice edges grow either by the freezing of seawater or by the transport of ice from other regions. Melting of the ice edge is induced by increased air and water temperatures, which is further enhanced by the breaking up of the ice floes and increased wave action in the MIZ. In some regions, such as the northern Weddell Sea, this decay phase can be considerably reduced if the transport of ice from the south through the Weddell Gyre is strong enough. The progression of ice edges, as the surface waters freeze or melt, can be very rapid. The speed of ice-edge growth and decay has been estimated as up to 2 km (1¹/₄ miles) per hour.

The equilibrium phase occurs when the ice edge reaches a northern boundary of warm water and the ice crossing the boundary melts. However, differences in the amount of ice or the rate that ice is moving northwards, combined with localised variations in air temperatures, can result in the ice edge oscillating between an advancing or retreating edge.

Above
The sea ice melts in the warmer waters at the ice-edge boundary.

Regional differences in ice-edge dynamics are evident in different regions of the Antarctic. For example, the Weddell Sea ice-edge undergoes a short advance period, a prolonged equilibrium phase, and a rapid decay phase due to the large volume of ice transported by the gyre. In other regions, the equilibrium phase can be very short at the maximum ice-edge extent, and ice advance and retreat are relatively long and nearly equal in duration.

Long-term trends in Antarctic sea-ice extent

In contrast to the slight decrease in the extent of Arctic sea ice measured over the past decades, there is a slight overall increase in the extent of Antarctic sea ice. It has been rising at a rather insignificant rate of approximately 0.4% per decade. In the Weddell Sea, Indian and Pacific Oceans the changes in extent are rather minor, but there are significant decreases of 8% per decade of ice in the Bellingshausen and Amundsen Seas and a corresponding increase of 7% per decade in the Ross Sea. These two sectors are adjacent to each other and it would appear that there is a movement of ice from the Amundsen Sea into the Ross Sea, as well as increased ice production in the latter.

Just as in the Arctic there is evidence that the extent of summer sea ice has decreased by about 3.7% per decade, although these decreases are considerably less than in the northern hemisphere. It is thought that the decreases are associated with earlier ice break up than in previous years. However, there is also evidence that overall ice begins to form earlier in autumn than before and hence the slight positive value for overall ice extent in the Southern Ocean since 1978.

These generalisations mask significant regional differences. There are measurements to show that the ice seasons in the eastern Ross Sea, far western Weddell Sea, and coastal regions of east Antarctic have shortened. In contrast, sea-ice seasons have lengthened in the western Ross Sea, Bellingshausen Sea and central Weddell Sea. In particular, the Antarctic Peninsula stands out as being a region with quite different temperature regimes from the rest of the continent. Over the past 19 years the region, especially the west coast of the peninsula, has experienced more rapid warming than any other part of the southern hemisphere.

El Niño and Antarctic sea ice

The El Niño Southern Oscillation (ENSO) is the largest climate oscillation on Earth to influence ocean currents and surface temperatures. El Niño is the term used to refer to unusually warm surface temperatures in the equatorial region of the Pacific. In contrast La Niña is the state when there are abnormally cold ocean surface temperatures in the region. During non-El Niño and non-La Niña conditions sea surface temperatures are about 5–8°C (9–14°F) warmer in the western tropical Pacific compared to the east, and the trade winds blow to the west across the region. The sea level is also higher in the western tropical Pacific because of the wind action.

During an El Niño period the sea surface temperatures increase significantly in the eastern tropical Pacific, and the trade winds either slacken or reverse direction, also moving less water from east to west, and greatly affecting the physical characteristics of the waters in the regions. However, the effects of El Niño are far more widespread than the Pacific Ocean, and weather patterns and ocean circulation patterns throughout the world are influenced by these events.

The Southern Oscillation part of ENSO refers to the east-west atmospheric circulation pattern characterised by rising air above Indonesia and the western Pacific and sinking air above the eastern Pacific. The strength of this circulation pattern is defined by the Southern Oscillation Index (SOI), which is a measure of the monthly differences in surface air pressure between Tahiti and Darwin. During an El Niño the surface air pressure is higher in the western tropical Pacific than in the eastern tropical Pacific and the SOI has a negative value. The reverse is true for La Niña periods.

Right
The Southern Oscillation Index (SOI) is a measure of the monthly differences in surface air pressure between Tahiti and Darwin, Australia. When there is a positive number there is a La-Niña (or ocean cooling), but when the number is negative there is an El-Niño (or ocean warming).

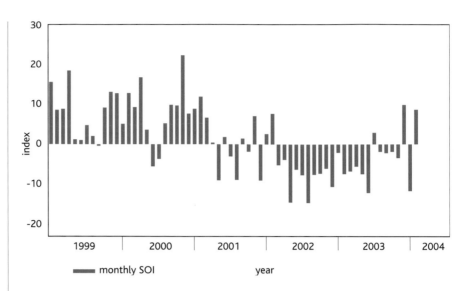

ENSO events generally happen every four to seven years and can last between one and two years. However, it seems as though in the 1980s and 1990s El Niño events were more frequent and lasted longer than previously recorded. There was a very protracted El Niño from 1990 to 1995 and exceptionally strong ENSO events in 1982/83 and 1997/98.

There is substantial evidence that the SOI is correlated closely to climate anomalies in certain sectors of the Southern Ocean, and that year-to-year variation in sea-ice cover in these regions is linked to recent ENSO events. When the SOI is in a positive phase there is generally a lowering of sea-level air pressure, surface air temperature and sea surface temperature in the Bellingshausen, Amundsen and Ross Seas, with the potential for greater ice growth. In contrast, during El Niño events (negative SOI) the reverse is true and reductions in the ice extent of these regions have been noted. In particular, the ENSO years of 1983, 1988, 1992 and 1998 show very good correlations with lower sea-ice extents, most notably in the Ross Sea. ENSO links to sea-ice distribution are not confined to these sectors but are also reported for cyclical sea-ice dynamics in other regions of the Southern Ocean.

Antarctic Circumpolar Wave

In recent years another major feature has been described for the Southern Ocean, the so-called Antarctic Circumpolar Wave (ACW). The wave appears to move around the continent in a clockwise direction. The wave takes about eight years to travel once around the Southern Ocean, and every three or four years any given longitude will experience a maximum in ice extent, whilst at the same time 90° of longitude to the east or west will experience minimum sea-ice extents. This 'wave' affects features not only at the ice edge but also deep into the pack ice. The combination of ENSO events and the progress of the ACW obviously account for a large proportion of the variations in sea-ice extent in the Southern Ocean. They are behind some of the striking year-to-year variations in regional sea-ice distribution.

Longer-term records of Antarctic sea-ice extent

One of the problems with trends drawn from satellite-based observations is that only 30 years of data are available. It is difficult to untangle long-term trends when there are cyclic events such as the ACW and episodic ENSO events that can distract from the long-term picture.

The positions of whales when they were caught or recorded during scientific expeditions from the early 1920s up to the 1960s, have been used to estimate the previous extents of the Antarctic summer ice edges. This is because the whaling for blue, fin and minke whales was largely restricted to ice-edge regions. Therefore, it was possible to make estimations of ice-edge extents before satellites were used, and compare these to the satellite data collected since the 1970s. Staggering conclusions that there had been a 25% drop in the extent of the summer sea ice in the Southern Ocean in just 20 years were widely reported, based on these comparisons. However, more recent analyses of such data indicate that whale catch data are not as reliable as first thought, and although there may be quite large regional differences in sea-ice extent, it is considered that the whaling data do not show such a massive reduction in Antarctic summer sea-ice extent between the 1950s to the 1970s.

Can gases in ice cores provide a better clue?

Despite some doubts in the whaling data, in 2003 there were reports from another source that also indicate a big reduction in the extent of sea ice in the Southern Ocean after 1950. Researchers measured concentrations of a chemical called methanesulfonic acid (MSA) in an ice core taken at Low Dome in eastern Antarctica. MSA is an oxidation product of dimethylsulphide (DMS), which is closely linked to the osmoregulation and temperature adaptations of sea-ice algae. When there is a large extent of sea ice, a lot of DMS and MSA will be released into the atmosphere, which will subsequently be incorporated into glacial ice as it forms. By measuring concentrations of MSA in the glacial ice core, researchers have been able to estimate sea-ice extent in a particular sector (80°E to 140°E) off the east Antarctic coast back to 1841. They estimate that sea ice extent in this region was rather constant between 1841 and 1950 and thereafter there was a sharp decrease of about 20%, not too dissimilar from the whaling record results. Naturally it is difficult to extend these results to the whole extent of sea ice surrounding the continent, but certainly combining MSA measurements from other cores from around the continent may well give an accurate measure of sea-ice extent over long periods of time.

MSA was also measured in an ice core recovered from the South Pole in 1995 (SP95). This core dated from 1487 to 1992 and the researchers working on it decided that the MSA records were related to the extent of sea ice in the Amundsen and Ross Seas. They concluded that the sea-ice extent in the Amundsen and Ross Seas was generally higher from 1800 to 1992 compared to the period from 1487 to 1800. They also demonstrated that ENSO-related atmospheric circulation systems (p. 59) have affected Antarctica for at least the last 500 years.

The advantage of using chemical properties as proxies for past climate conditions in these long ice cores is that environmental records covering thousands of years are locked up in the ice. The Low Dome ice core extends back 80,000 years, and so it is possible that proxies such as MSA in these cores may provide estimates far back into geological time. The only down side is that for the comparisons to be reliable much research has to be conducted on how the chosen proxy may change in nature during long time periods and, of course, to determine beyond doubt that nothing else regulates the proxy except the environmental parameter in question.

The Baltic, Caspian and White Seas

Although the Arctic and Antarctic are by far the most extensive and arguably the most important bodies of pack ice, there are smaller bodies of saline water that are influenced by ice. Even temperate regions experience sea ice from time to time, especially in enclosed water bodies such as lagoons and sheltered bays. To discuss all of these is beyond the scope of this book, but briefly three large water bodies, the Baltic, Caspian and White Seas, will be introduced. Sea ice is a major feature in each of these seas, causing seasonal difficulties in commercial shipping traffic and major ecosystem transformations of similar magnitudes to those experienced in the Arctic and Southern Oceans. However, these three smaller seas are all characterised by water salinities much lower than typical ocean values.

The Baltic Sea

The Baltic Sea is a semi-enclosed sea that covers an area of approximately 415,000 km² (160,000 sq miles). It has an average depth of 55 m (180 ft) and a deepest point of 459 m (1506 ft) at Lansort Deep. There is a large freshwater input into the sea via rivers, rain and snow. The Baltic is the largest brackish water basin in the world, and the saline input is maintained by periodic inflows of Atlantic waters through the Danish Straits. These occur during irregular events called salinity pulses that are created by strong winds pressing surface waters eastward through the Danish Straits.

The average salinity in the Baltic is lower than 10 and there is a surface water salinity gradient of approximately 20 at the Danish Straits, decreasing to almost 0 in the extreme northern Baltic or close to St Petersburg in the east. Because of the low salinity, the water freezes from 0 to –0.2°C (32 to 31.6°F) in the Bothnian Bay, but at –0.5°C (31.1°F) in the more saline Baltic proper where salinities are around 5 to 6.

Sea ice is present in the Baltic Sea generally from November to June, with a maximum ice extent between January and March. On average the ice covers an area of about 200,000 km² (77,220 sq miles), although during mild years this can be as low as 52,000 km² (20,000 sq miles). In very severe years over three quarters of the Baltic can be covered by sea ice, albeit for a short time. There is good evidence that the severity of the ice cover and the length of the ice season is related to AO/NAO events (p. 52) as in the Arctic Ocean.

Above
The Gulf of Finland in the Baltic Sea completely covered by sea ice in January 2003. The Gulf is an important shipping route bound by Finland to the north, in the south by Estonia, and to the east by Russia. At the far eastern end of the Gulf lies St Petersburg.

There are long-term records dating back to 1529 that describe the severity and extent of the ice seasons in the Baltic port of Riga, Estonia. There is some evidence that there is a decreasing trend of about two days per century in the length of severe winters, although no trends have been spotted for average or mild winter seasons. Interestingly there was an almost even split in the percentages of mild, average and severe winters during the period 1529 to 1990, based on these records and other historical documents.

Ice formation begins in the lowest salinity waters in the northernmost parts of the Bothnian Bay and in the easternmost parts of the Gulf of Finland progressing to the Bothnian Sea, the Archipelago Sea and the rest of the Gulf of Finland, and then into the northern part of the Baltic Sea proper. In severe winters sea ice expands further to cover the Danish Straits and the Baltic Sea proper.

Ice melts from the south to the north and generally the northern Baltic Sea is ice free in April. Ice persists until late May/June in the Bothnian Bay, and so although the ice may only last for periods of days up to a month in the Baltic Sea proper, it may last up to six months or more in the northern Bothnian Bay.

In shallow waters up to 15 m (50 ft) deep, and in between the coastal islands and archipelagos, land-fast ice develops and forms a rather stable platform. Pack ice covers deeper waters, and in parts of the Baltic severe deformation can cause large ridges to form that are many metres thick.

The White Sea

The White Sea is an almost enclosed body of water covering an area of about 90,000 km² (35,000 sq miles) in northern Russia. The sea is formed from three large bays, the Dvina, Onega and Kandalaska Bays, that are connected to the north through the narrow Gorlo Strait to the Barents Sea. It is a rather shallow sea with a mean depth of 89 m (291 ft) and a maximum depth of about 300 m (984 ft). The White Sea has a more continental climate than the Barents Sea with warmer summers and harsher winters. There are large inflows of river water into the sea through large rivers such as the Kem, Onega and Dvina. The exchange of water with the Barents Sea is rather restricted and so with the large freshwater input the salinities of the White Sea are reduced compared to those of the Barents Sea. The salinities of surface waters range from 25 to 28, and these overlay deeper waters that collect in the basins that may have salinities up to 30. Clearly the regions of the large river inputs are places of considerably lowered salinities and therefore regions where freezing of water takes place rapidly.

Considerable amounts of sea ice forms in the White Sea from the end of November through to the end of April, although in extreme years ice may persist for up to seven months. Generally there is a region of well-developed

Above
Ice covered Baltic shoreline.

Opposite
Large pressure ridges tower over a scientist in the Baltic pack ice.

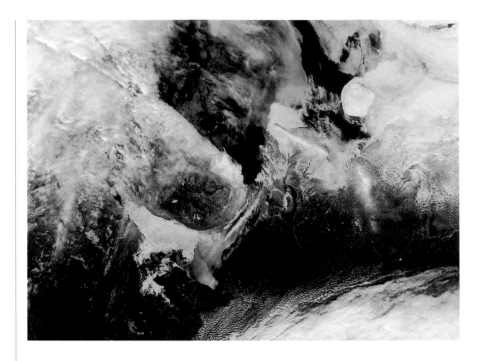

land-fast ice that may extend for several kilometres towards open waters where more open pack ice is encountered. The coastline of many parts of the White Sea comprises narrow fjord-like inlets, up to several kilometres wide. These features are frequently covered by continuous sheets of land-fast ice.

Sea ice can reach thicknesses of over a metre, although just as in other pack-ice regions deformation of ice floes and the formation of pressure ridges can result in sea ice several metres thick. The general circulation of the White Sea results in the sea ice being carried out to the Barents Sea. No sea ice persists through the summer, since water temperatures rise to 15°C (59°F) in open waters and up to 20°C (68°F) in sheltered coastal inlets and bights.

The Caspian Sea

The Caspian Sea is the largest enclosed body of water on Earth, covering an area of approximately 386,000 km² (149,000 sq miles). The Caspian lies approximately 30 m (98 ft) below sea level. The sea is surrounded by Azerbaijan, Iran, Kazakhstan, Russia and Turkmenistan, a meeting point between the Middle East, Europe and Asia. It is the remnant of an ancient ocean Tethis that used to connect the Atlantic and Pacific Oceans about 50–60 million years ago. This accounts for why the sea is still saline. The northern part of the Caspian, which covers up to 25% of the total area is very shallow with average depths less than 10 m (33 ft). In the south Caspian, depths reach over 1000 m (3300 ft), and the average depth of the whole sea is about 200 m (656 ft).

A number of very large rivers flow into the northern and western part of the Caspian, the most notable being the river Volga, which contributes about 70% of the total river water entering the sea. The huge volumes of river water result in the salinities of the northern Caspian falling below the average of 12 down to values less than 5 close to the entrance of the Volga.

Almost all of the northern Caspian is covered in ice between December and March. In general the ice reaches a thickness of 40–50 cm (16–20 in), and is up to 70 cm (28 in) thick in the northernmost regions. Ice formation starts in the lowest salinity waters, gradually moving southwards towards the central northern Caspian. The southerly limit of the pack ice rarely extends further than the part of the Caspian that is more than 12 m (40 ft) deep.

The Caspian Sea has major oil and gas reserves that are only just beginning to be fully developed. Oil reserves for the entire Caspian are estimated to be comparable to those of the USA and North Sea, and the natural gas reserves even larger. The Kashagen oilfield could prove to be the world's fifth largest oilfield, and the rigs working in these shallow seas have to be able to withstand the seasonal ice fields.

One of the most striking features of the northern Caspian is the extremes of temperature experienced over the year. Sea ice and freezing air temperatures below –20°C (–4°F) are characteristic of winter, but in summer the water temperature rises up to 31°C (88°F) in the shallow waters. The temperature ranges in these waters are therefore very much greater than those found in polar regions. The extreme range of temperatures, from tropical to polar, make these waters arguably far more hostile places for organisms to cope with than regions where temperatures are more consistently cold.

Above
The shallow part of the northern Caspian Sea becomes covered by sea ice including the large river entering the sea on the top left-hand side of the image.

Life within a block of ice

4

GREY, WHITE AND BLACK ARE ALL COLOURS USED TO DESCRIBE SEA ICE, BUT THE COLOUR MOST COMMONLY ASSOCIATED WITH PACK ICE IS A COFFEE-BROWN STAINING ON THE UNDERSIDES OF MANY ICE FLOES. THIS BROWN ICE GIVES THE FIRST HINTS THAT THE PACK ICE IS NOT A FROZEN WHITE DESERT DEVOID OF LIFE, BUT THAT RATHER THE ICE IS A HABITAT IN WHICH A DIVERSE AND RICH BIOLOGY DEVELOPS.

THE BROWN COLOUR is from photosynthetic pigments contained within microscopic algae living in the ice. These ice-bound organisms have a pivotal role in maintaining the polar ecosystems as we know them, even having vital consequences for polar mammals and birds. Ever since the first explorers ventured into pack-ice zones, high concentrations of brown ice and considerable biological activity in around those floes has been noted in expedition reports and ships' logs. Fridtjof Nansen sums up some of his excitement of observing life within the ice as he sat over his microscope in the ice-bound *Fram* from 1893 to 1896: "And these are unicellular pieces of slime that live by the million in pools on very nearly every ice floe all over this endless sea of ice, which we like to call a place of death! Mother Earth has a strange ability to produce life everywhere. Even this ice is fertile ground to her."

However, one of the first people to describe sea-ice organisms was Sir Joseph Dalton Hooker in his *Flora Antarctica*, published in 1847. In this book he described many of his findings during the pioneering voyages of HMS *Erebus* and HMS *Terror* from 1839 to 1843, commanded by James Clark Ross. This publication followed the descriptions of dense growths of algae within Arctic pack ice by the eminent German protozoologist

Opposite and below
Sea ice is often coloured brown, indicating that the ice is inhabited by rich biological assemblages.

Diatoms are the most conspicuous components of the sea-ice biology.

From top:
Original drawings of Antarctic diatoms by Sir Joseph Hooker during the Sir James Ross Antarctic Expedition 1839-43.

An uncannily similar picture of diatoms from the Weddell Sea ice.

A diatom *Nitzschia stellata* viewed under a microscope (this is a common diatom in Antarctic sea ice).

A variety of different diatom species isolated from sea ice in McMurdo Sound, Antarctica.

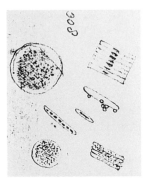

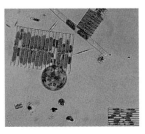

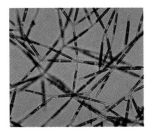

C.G. Ehrenberg in 1841 and 1844. Almost a decade later P.C. Sutherland went to the Arctic as part of a team searching for the lost ships *Erebus* and *Terror* (this time used for the ill-fated Franklin expedition to find the Northwest Passage). He reported "greenish slimy-looking substances" on the underside of Arctic ice floes. When he examined the slime under a microscope he found "minute vegetable forms of exquisite beauty".

The organisms that these eminent scientists were describing were diatoms: microscopic unicellular organisms found in rivers, lakes and oceans around the world. When Ehrenberg and Hooker wrote about the ice diatoms, they were largely considered to be animals. We now know them to be microalgae, aquatic unicellular organisms that photosynthesise like plants. Diatoms require sunlight and nutrients such as oxygen, carbon dioxide, nitrogen, phosphorus, and a whole host of other trace elements to grow. They replicate themselves by dividing into two cells, and under ideal conditions they form blooms where division is so rapid and extensive that the water in which they are living can be turned from crystal clear to a murky brown in just a few days. Diatoms are in turn a food source for grazing planktonic animals such as amphipods, copepods and euphausiids (krill).

Other microbes in the ice

Although it is the pigments contained within diatoms that mostly discolour the ice, there are a multitude of other micro-organisms that thrive within the sea-ice matrix. These include viruses, bacteria, other types of algae and protozoa. The study of sea-ice ecology has concentrated on the study of diatoms, foraminiferans and bacteria. This does not necessarily mean that they play the greatest roles within the ice, but is probably more to do with the fact that they are easier to sample and work with than some of the other groups.

The micro-organisms form complex metabolic networks, in which some of the organisms such as the bacteria and diatoms are food for grazing protozoa such as the heterotrophic dinoflagellates, ciliates, amoebae and nanoflagellates.

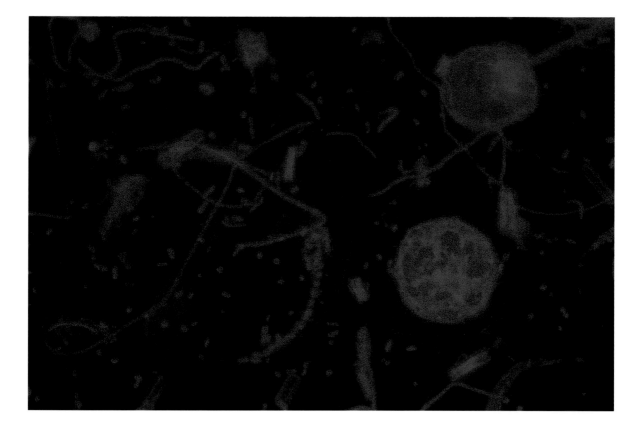

The interactions and deciphering of 'what is eating what' is a complex task, especially since we do not know many of the organisms that are actually there, let alone their life cycles and food preferences.

Larger organisms in the ice

This complex network of microbial organisms is a rich food source for larger organisms. They can be divided into two groups of organisms: those that are found living within the ice, and those that are closely associated with the edges of ice floes, grazing on other organisms that live on the ice surfaces.

The larger organisms complete a larger food web that is ultimately based on the bacteria and algae growing within the ice. However, the larger organisms such as the fish, amphipods and krill, not only eat bacterial films and algae, but they also eat other animals. These larger metazoans are in turn a food source for squid, some species of seal, whales and seabirds.

Where does the ice biology come from?

Micro-organisms are introduced into the ice as it is forming. As the frazil ice crystals float towards the surface (p. 17), they trap particles in the water

Top
The orange dots are sea-ice bacteria that have been stained with specialised fluorescent dye.

Above
Foraminiferans commonly reach high numbers in Antarctic sea ice, and are only occasionally found in Arctic ice.

Right
The variety of some of the organisms living within and at the edges of Antarctic ice floes. Note the illustrations are not drawn to scale: a, diatoms; b, flagellates; c, foraminiferans; d, ciliates; e, turbellarians; f, crustaceans; h, g, copepods; i, crustacean larvae; k, krill and l, young fish.

Below right
The amphipod *Gammarus wilkitzkii* grazes on the underside of Arctic ice floes.

between the crystals. Therefore any planktonic organisms present in the water will effectively be concentrated into the grease ice that accumulates on the surface of the ocean.

Organisms that are not able to swim away from this ice mass become trapped and form the basis of the initial sea-ice biology. The ice organisms are mainly planktonic organisms, simply caught up into the ice matrix. Not all the organisms will survive in the ice, of course, but when the ice melts, its contents are released back into the water and the ice biota will return to being plankton once more. The entrapment in the ice is a random process, and so it is wrong to talk about a 'community' at this stage. The term assemblage seems more appropriate.

Tolerant or thriving at freezing temperatures?

Clearly the ice habitats are considerably colder than the surrounding waters, and not all of the plankton can survive at temperatures below freezing. Organisms that require low temperatures are called psychrophiles. Strictly these organisms have optimal temperatures for growth below 15°C (59°F) and minimum growth temperatures well below 0°C (32°F). Organisms that can grow at or below 0°C, but have higher optimum growth temperatures at 20–40°C (68–104°F) are termed psychrotolerant.

Many of the organisms living in polar waters are psychrotolerant, and when they are caught up into the freezing ice matrix their growth is severely inhibited and in many cases ceases. In contrast, psychrophilic species thrive in the ice

Left
As new ice forms, the organisms harvested out of the water by rising ice crystals bloom in the grease ice as it changes to pancakes on the ocean's surface.

Above
Only a fraction of the organisms found in polar waters are capable of tolerating 'life within the ice'.

habitats. Some species of bacteria are known to respire and synthesise proteins down to temperatures of –20°C (–4°F) where there is hardly any water left in the brine channels. Therefore, although the sea-ice biology is recruited from the plankton, it is only a proportion of these organisms that will actually grow and thrive within the ice.

Many ice algal and dinoflagellate species have been shown to continue growing and dividing at –8°C (18°F), and a corresponding salinity of 145. However, it takes these organisms up to 60 days to grow and divide to produce the next generation under these conditions, very much longer than the generation times of three days at –4°C (25°F) and about a day at 0°C (32°F).

PUFA production at low temperatures

The critical requirement for survival at low temperatures is that the cell membranes, which are principally composed of lipids (fats), remain in a liquid phase so that they can function properly. The effect is similar to a block of butter, which is pliable when warm, but becomes more solid when cooled, and completely solid when put in the freezer.

The degree of fluidity depends on the membrane's fatty acid composition: the greater the amount of unsaturated fatty acids and polyunsaturated fatty acids (PUFAs) the higher the fluidity of the membrane. Increased concentrations of PUFAs in membranes have been shown to be central to low-temperature adaptation in both sea-ice algae and bacteria. Carrying on the butter analogy, butter that is high in saturated fatty acids solidifies at much warmer temperatures than margarine, which is mainly made up from unsaturated fatty acids. In general, the membranes of psychrophiles have a higher proportion of unsaturated fatty acids.

Most studies of lipid and membrane metabolism have been conducted on psychrophilic bacteria. However, in more recent years the regulation of lipid composition in the cell membranes of sea-ice diatoms has also been shown to be essential for cellular metabolism at low temperatures, low irradiances and with limited inorganic nitrogen nutrients.

Enzymes and low temperatures

Enzymes are proteins that act as catalysts in chemical reactions: the rate of a process can be increased by having more enzymes or changing the efficiency of an enzyme. Psychrophilic organisms are characterised by enzymes with high catalytic efficiencies at low temperatures, and some bacteria respond to a lowering of temperature by producing forms of enzymes that have a greater efficiency than the same enzymes produced at warmer temperatures. Another strategy is simply to increase the concentrations of enzymes with decreasing temperature, thereby counterbalancing reductions in efficiency.

Tolerating high and low salinities

All organisms living within or at the periphery of ice floes experience deviations from normal seawater salinities. Decreasing temperatures within the ice result in brine salinities greater than 200, whereas when ice melts salinities close to freshwater (salinity of 0) will be encountered. Sometimes the changes in salinity can be rapid, and so the organisms must be able to adjust quickly to a wide range of osmotic changes. This is largely brought about by changes in concentrations of ions and solutes within the cells.

In normal seawater the cells of algae and bacteria are in osmotic balance with the water outside of the cells. When the salinity of the external water goes down, the induced osmotic imbalance results in water being taken up by the cells, and the amount of water taken up is in proportion to the reduction in external salt concentration. The cells combat this uptake of water by lowering the cellular concentrations of ions such as potassium, sodium and chloride, as well as sugars and organic compounds. This restores the osmotic equilibrium to one close to that before the salinity had been reduced. If the organisms don't do this they will take in water until the cells burst.

When the salinity of the external water increases, the reverse process takes place. The higher concentration of external salts causes water to be lost from the cells, again in direct proportion to the increase in salinity. The cells have to restore the osmotic balance by taking up ions, or producing greater cellular concentrations of sugars and organic compounds. These sugars and organic compounds are many and varied, including glucose, sucrose, proline and dimethylsulphonioproprionate (DMSP). Low temperatures also initiate the production of many of these compounds, and so they often have a dual role in protection at low temperatures and salinity tolerance.

DMSP is a precursor to the volatile gas dimethylsulphide, DMS. This is oxidised in the atmosphere to a range of substances that act as cloud condensation nuclei from which clouds are formed. Therefore DMSP produced to combat salinity and low-temperature stress by sea-ice algae can ultimately play an important role in localised and large-scale climate regulation. It is DMS that together with ozone creates that wonderful 'seaside aroma'.

Clearly, melt ponds on the surface of ice floes have reduced salinities as do regions in the Arctic where large rivers discharge into shallow seas. Quite often these waters and ponds have a large proportion of freshwater species of algae and bacteria growing in them. If these species are incorporated into high-salinity brines when ice refreezes, or are washed into the surrounding seawater they die, since in general they are less tolerant to salinity change. *Phaeocystis* species are well known DMSP/DMS producers in waters around the world, and *Phaeocystis* blooms in the ice are associated with very high DMS release. Sometimes the stench of DMS being released from the ice can be overwhelming as an icebreaker moves through rotten summer ice floes full of *Phaeocystis*.

Other volatile gases released from ice

DMS is not the only volatile gas to be emitted from marine algae with consequences for atmospheric chemistry. Reactive halogen species contribute significantly to the destruction of ozone in the polar stratosphere as well as the underlying troposphere. Both Arctic and Antarctic sea-ice algae also produce significant quantities of bromoform, dibromomethane, bromochloromethanes and methyl bromide. The high levels of these gases produced from sea ice contribute to the destruction of ozone in the polar atmospheric systems and may be of a similar magnitude, on a global scale, to amounts of these compounds released from anthropogenic sources and the other major producers of such chemicals: the seaweeds.

How does the biology survive in the ice?

Within any single ice floe there are several regions in which the ice organisms are found. Primarily the organisms grow within the brine channels and pores in the ice, although dense growths of bacteria and algae can accumulate within water trapped between looser agglomerations of ice crystals such as between ice platelets (p. 21). A major factor governing whether or not an organism survives is the space available.

Below
Ice diatoms within a brine channel in Arctic sea ice.

Living in gradients

There are gradients of physical factors within the ice that clearly govern the type of habitat for life to proceed. As described previously (p. 23) the volume of the brine channel system is governed by the temperature and age of the ice. Typically the surface of the ice is colder than the ice at the bottom of a floe, which is always at or just below the freezing point of seawater. This sets up a gradient of temperature between the surface and bottom of the floe, the steepness of the gradient being determined by the air temperature. This means that the size and volume of the channels and pores in the uppermost parts of the ice are far smaller than those towards the increasingly warmer ice, moving downwards to the ice/water interface. A consequence of this is that there is more space for organisms to live within the warmer ice at the bottom compared to the surface layers of ice. Likewise as ice gets colder, the salinities of the brines increase and brine drainage processes are enhanced. Therefore the highest brine salinities are found higher in the ice.

The other factor that is fundamental for photosynthetic organisms living in ice is the amount of light available. The thicker the ice, the less light there will be penetrating to the underside of an ice floe. It is not just the quantity of light that is affected, but also the quality. For instance, ultraviolet wavelengths of light are more effectively absorbed than others. Evidently the number of pores, channels or gas bubbles will also greatly influence the light field within the ice. These features will tend to refract and reflect light giving the ice very different optical properties compared to a simple, solid slab of ice.

The depth of snow cover overlying the ice is key in determining the extent of the temperature and salinity gradients, as well as affecting light levels. This is because snow is a good reflector of light as well as being a good insulator. Therefore ice floes with no snow cover will be far more exposed to changes in temperature and salinity than floes that have a thick snow cover. Similarly, a thick cover of snow will effectively block out the light.

Changing habitats

Changes in temperature can occur rapidly even in the space of 24 hours. Naturally this will be greatest in the surface ice layers. Such changes affect the brine volumes and salinities in the ice, which in turn have a significant influence on the biology living within the ice.

Even the light regime has the possibility of changing on short timescales, since snow cover can be rapidly washed or blown away, increasing the light penetrating the ice. On the other hand if snow is blown into drifts or piled up, naturally this will greatly reduce the light available for photosynthesis. Flooded wet snow provides less of a barrier to light than dry snow that reflects a greater percentage of the incident light.

Right
The highest densities of biology occur on the peripheries of ice floes, especially on the bottom surfaces.

Below right
Illustration to show the different areas of ice in which biological assemblages develop.

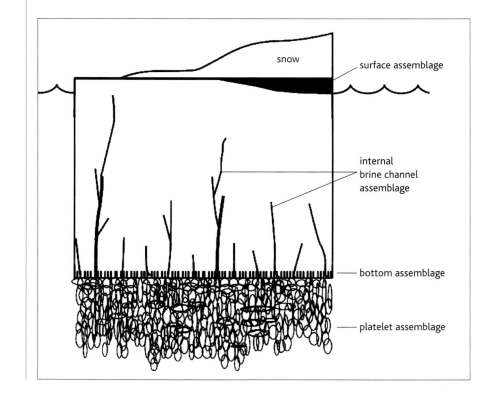

Where is the ideal place to live?

Even though light levels can be low, it is in the lowest few centimetres that the highest concentrations of biology are often found. The bottom of an ice floe, within the skeletal growing layer, is always at or just below the freezing point of seawater. The salinities of the brines are therefore closer to seawater values, and the brine channels and pores are large – up to a few millimetres in diameter. The organisims living in the lowest regions of the ice are the so-called 'bottom' sea-ice assemblages.

Moving upwards into the interior of a floe, temperatures and brine salinities become more and more distinctive from those in the bottom ice region. In general, the ice becomes colder and so brine channels and pore diameters become smaller and the interconnections between channels and pores become fewer. The brine salinities are also generally higher. Despite the increase in potential stress factors that combine to inhibit biological activity, assemblages of organisms are frequently encountered within the interior of ice floes and sometimes these can reach high biomass.

Ice fields are dynamic places where terrific stresses and strains can be placed upon the ice. Floes may fracture and crack with the result that surrounding

Below
Ice, darkly stained by diatoms in the brine channels, absorbs more heat than clear ice, and this can result in the melting ice having a honeycomb texture.

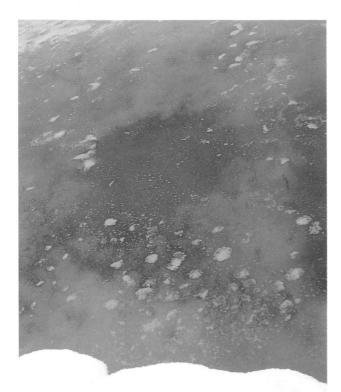

Above
Melt ponds can have dense growths of sea-ice diatoms, on the ice surfaces as well as within the pond water.

seawater is able to penetrate the interior of floes, causing localised melting and also an input of less stressful salinity conditions that are conducive to growth of algae, bacteria and their grazers.

Typically the uppermost layers of ice are relatively devoid of life, only becoming a habitat when ice starts to melt, or special surface features are formed. This is despite the fact that on any ice floe it is in the surface layers that light levels are at their highest, and therefore optimal for photosynthesis to take place.

Subsurface gap layers – an ideal ice habitat?

As discussed earlier, surface flooding, snow-ice formation and melting processes can result in subsurface gap layers and voids forming. These are typically permeated by seawater, which is regularly exchanged by waves pumping water through the voids. Organisms growing within these layers are not only exposed to high light levels, which induce maximum growth of photosynthetic organisms, but they are also exposed to less stressful temperatures and salinities compared to those found in 'normal' sea ice. Because of this, a high biomass of sea-ice organisms can be found within these specialised ice habitats. When high accumulations of ice diatoms are present in these voids, they absorb light to such an extent that they may enhance the rates at which surface ice melts.

Surface melt ponds may also contain high accumulations of living material. Although these are high light habitats, and therefore sites where algal growth should be high, they generally contain very low salinity water. In many ways these melt ponds can be considered as contained freshwater lenses that are floating on the surface of a saline ocean. Organisms found in melt ponds must be able to survive salinities very much lower than seawater values. In the Arctic, where summer melt ponds are a very distinctive feature, many of the species identified in melt ponds are in fact species more commonly found in rivers and lakes. The micro-organisms reach the sea-ice ponds via river runoff, or they may be carried on the wind. Some are transported to the melt ponds on the feathers and feet of seabirds.

Although there is layer of ice between the pondwater and the ocean, this is often highly porous and there may well be interconnections between the two types of water. This allows organisms that can cope with the osmotic stresses to move from the ocean into the melt ponds to feed.

Above
'Clouds' of platelet ice under the overlying ice are habitats where dense growths of algae occur.

Platelet ice – a unique ice habitat

The ice platelets that accumulate under land-fast ice in the Antarctic support dense assemblages of ice organisms. These form semi-rigid 'clouds' of ice platelets that trap interstitial seawater between the ice crystals. Depending on the depth of the platelet layer, typically 1–5 m (3–16 ft) thick, and the turbulence of the underlying water, the interstitial water is periodically exchanged in the platelet layers. Since these layers are under sea-ice floes, light is often limiting, but the organisms trapped within the platelet layer do not experience salinities significantly different from seawater. The temperature of the interstitial water is also just above freezing temperatures, and so besides the light limitation, platelet layers are a sea-ice habitat where growth conditions are good. In fact the highest biomass accumulations of sea-ice algae have been recorded in platelet layers.

Sometimes the overlying fast ice cracks apart because of tidal water movements, and the platelets rise through the crack to come to the surface of the ice. These platelet-filled cracks can support very high concentrations of algae, because moving from a 'low light' to a 'high light' environment boosts the algal growth rates.

Patchiness of life in the ice

It is a well-known feature of any study of aquatic systems that organisms and chemical constituents are not distributed throughout a water body in a uniform pattern. Such 'patchiness' is also a feature of sea-ice assemblages, which is not surprising considering that the organisms are recruited from the water in the first place and must depend on small-scale variations in the physical characteristics of the sea ice, such as the non-uniform distribution of brine channels. Just as it is possible to take a water sample from one side of a boat, and then get a totally different sample from the other side, it is possible to take an ice core that is full of biology and then take another one just a few centimetres away and find that it is completely different. The thickness of ice can also vary greatly within small areas, with obvious effects on light regime and quality of habitat.

The underside of an ice floe is seldom a flat surface, and extensions of the skeletal layer protrude to different degrees into the underlying water. The

Right
The variation of ice-core lengths taken within a small area of Arctic sea ice illustrate the heterogeneity of sea ice.

combination of the uneven surface and varying water currents under the ice can lead to accumulations of biology in pits in the ice surface or against the edges of ice protrusions.

Of course, the greatest factor causing unevenness of the underside of the ice is rafting of ice and formation of pressure ridges that can have large keels, extending many metres into the underlying water. These form a series of cracks and crevices, even small caves, which provide surfaces of very different qualities for colonisation by biological assemblages. The distribution of organisms in such areas is very heterogeneous, and consequently very difficult to sample effectively.

Below
Sea-ice stalactites are hollow, tapering structures up to a metre long. As brine drains downward into the sea, ice forms around the draining cold brine forming a hollow ice stalactite.

This patchiness is one of the greatest brain-teasers facing ecologists who are interested in the numbers and diversity of living organisms contained within sea ice, and how this varies from region to region. In the open ocean we have satellites that are able to measure the distribution of phytoplankton in the waters by measuring the colour of the water and the photosynthetic chlorophyll pigments. High resolution images of phytoplankton distribution can be compiled that show both spatial and temporal developments from day to day, thereby giving a measure of the variability of phytoplankton distribution.

New technology to capture patchy distributions

However, it is not possible to use satellite technology to get images of the biology contained within or below sea ice, since the ice effectively blocks the sensors. The only reasonable alternative is to look at the ice from below. In the past few years remotely operated vehicles (ROVs) have been used to examine the distribution of sea-ice biology. These vehicles are tethered by a cable to a 'driver' on board the ship and can be manoeuvred under the ice. Typically they have still and video cameras giving the operators 'real time' pictures of the distribution of organisms and biological accumulations on the underside of the ice. The method is limited, however, in that an ROV cannot operate at distances beyond the length of its cable, which is typically a few hundred metres long.

In recent years there have been the first attempts to use autonomous underwater vehicles (AUVs) to map the distribution of ice algae growing on the undersides of the ice. The advantage of AUVs over ROVs is that they can be programmed to travel several hundred kilometres in any one run, thereby enabling the study of a greater spatial distribution of sea-ice biology. These vehicles do not use cameras, but have specialised sensors that measure the fluorescence of the chlorophyll contained within the algae when they are illuminated by flashes of light sent out from the AUV. The higher the fluorescence, the greater the concentration of algal cells.

The disadvantage of AUVs is that they are pre-programmed to set courses, and as such cannot operate too close to the underside of the ice for fear of hitting pressure-ridge keels. Unlike the ROV, where real-time observations can be made, the AUV collects data to be downloaded on return to the ship, although potentially it could be programmed

Below
Remotely operated vehicles (ROVs) equipped with powerful video and camera equipment are used to survey the underside of ice floes and the seafloor.

Left
Autosub is an autonomous
underwater vehicle that can be
equipped with different sensors
and be deployed for long
periods far under the sea ice.

Below
Autosub just before
diving, with a curious whale.

to surface at set intervals and send the data via satellite telecommunications to the operators. However, at present the development of AUV surveys seems to be the only realistic means by which we will determine large-scale distribution patterns of bottom ice algae assemblages.

Clearly both of these methodologies can only be used to map the distribution of bottom biological assemblages. We still lack the ability to record large-scale distributions of interior and surface assemblages.

Microbiology inside the ice

5

THE SPACE INSIDE THE BRINE CHANNEL SYSTEM DICTATES THE SIZE OF THE ORGANISMS FOUND WITHIN IT. SO, IN MUCH OF THE ICE BELOW −5°C (23°F), THE RESIDENTS ARE MICRO-ORGANISMS THAT CAN BE SEEN ONLY WITH A MICROSCOPE. THEY INCLUDE A RANGE OF VIRUSES, BACTERIA, ALGAE AND PROTOZOA THAT FORM COMPLEX MICROBIAL FOOD WEBS, AND ARE THEMSELVES A RICH FOOD SOURCE FOR LARGER ORGANISMS.

THE BASIS of the oceans' food webs is the unicellular, mostly microscopic algae that make up the phytoplankton. There are many different forms, ranging in size from less than a micrometre to 0.5 mm. Numerous algal species are found in sea ice, although it is the diatoms that have been best studied. Approximately 200 species of diatoms have been described in Arctic ice and 100 species have been reported from Antarctic ice. Photosynthetic dinoflagellates and nanoflagellates are the other major photosynthesising organisms within sea ice, and in certain types of ice these can reach concentrations that even dwarf the biomass of diatoms.

Diatoms are hard nuts to crack

One of the reasons that diatoms are the best studied of the sea-ice algae is because they have a silica-strengthened cell wall (frustule) that makes them resilient to physical damage and therefore easy to sample. Many of the other algae such as flagellates are considerably more delicate, and traditional sampling methods tend to damage them to such an extent that they are overlooked.

The frustule is made of transparent silicate, which is basically glass. However, diatoms are not fragile, easily shattered crystal cases. On the contrary, miniaturised crash tests combined with computer simulations on 'virtual' diatom frustules have shown that they can withstand pressures equivalent of up to 700 tonnes per sq metre before they crack.

Evidently the intricate architecture of diatom frustules has evolved as a tough mechanical protection. In turn, as part of an evolutionary arms race,

Opposite
Under the snow and within the ice a rich fauna and flora can develop.

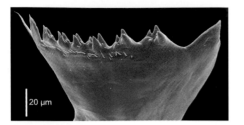

Above
The mouthparts of some crustaceans are specially strengthened to crack open tough diatom frustules.

copepods and krill, which rely on diatoms as a food source, have developed specialised tools to crack the diatoms open. These include hard, silica-edged mandibles and gizzards lined with hard crushing 'teeth'. But even these are not enough, and massive muscles are needed to operate these crushers. The smaller zooplankton species have no chance of cracking open a diatom and have to graze on less demanding food sources.

Growing on crystals and in slime

If you were to place an open glass of water or bottle on a windowsill, within a few days the inside surface of the glass would be covered by a film of green microalgae. Many species of microalgae grow well on wet surfaces (even on wet trees, walls or garden furniture). Included among them are many of the species of diatoms, which grow within the brine channels of sea ice as well as on the ice crystal surfaces. Pennate diatoms – the most common sort of diatom in ice – are even able to move along ice crystal surfaces, and so are able to move from one part of an ice floe to another. In fact, it was the ability of diatoms to move that made many of the early scientists think that they were animals.

Some diatoms have been shown to produce polysaccharide glues that enable them to stick to surfaces. Others produce films of polysaccharides to create a slime matrix that surrounds groups of diatoms embedding them onto a surface such as an ice surface. These so-called biofilms are also produced by bacteria, and often bacteria and algae are embedded into the same biofilm. Such biofilms form on just about every surface in aquatic systems, including rocks, sand grains, seaweeds, ships' hulls and marine animals. Therefore, it is not surprising that they form on exposed ice crystals as well.

The polysaccharide material released by diatoms, called extracellular polysaccharide substances (EPS), has recently been described as having a cryoprotectant role for Arctic sea-ice diatoms and possibly bacteria. It is thought that large quantities of EPS within the brine channel system may greatly alter brine fluid viscosity, and rather than a liquid-filled labyrinth it is possibly more like a weak gel-filled system.

Researchers working with these topics may well turn for advice to research conducted on soils, sediments and even more unlikely subjects such as rocks. The reason for this is that porous substances such as these often contain very well developed microbial populations made up of algae, bacteria and fungi, and there may be intriguing similarities in the ecological interactions of organisms living in EPS-rich environments.

Ice-pitting substances

As well as polysaccharides, some species of ice diatoms have been shown to release 'ice active substances'. These mostly unidentified compounds cause

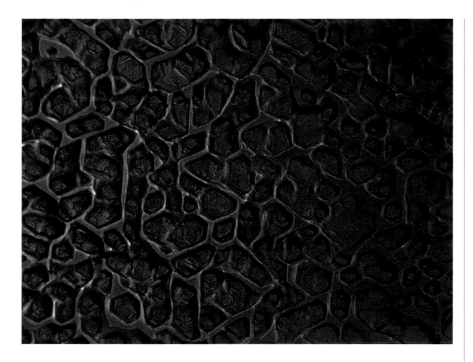

Left
The surface of an ice crystal that has been 'pitted' by ice-active substances released by ice diatoms.

pitting and deformities on the surfaces of growing ice crystals. Research into these substances is still at a formative stage, but there is strong evidence that they protect the cells from freezing damage, even though the precise cryoprotection mechanisms are unclear. Experiments with 'crude' extracts of the substances show that they bind to ice-crystal faces and are incorporated into the ice lattice, thereby affecting the ice-crystal formation and consolidation processes.

Algal species diversity in the ice

Despite the large diversity of photosynthetic organisms identified from sea ice, generally only a few species dominate. As first-year ice grows, the species diversity gradually decreases from the highly diverse inoculum to a small group of smallest species. At very high algal densities, the whole assemblage can often just consist of one species. It was thought that it was the small pore and brine channel size that selected for these small species, but recent research has proposed that it is their physiological capabilities that result in their selection. In the Antarctic two small diatoms *Fragilariopsis cylindrus* and *Fragilariopsis curta* are particularly ubiquitous in sea ice, and whereas some species of diatom seem to be described in localised regions, these two are reported in most sea-ice studies. Where they occur they also tend to be the main species present.

Dense photosynthetic dinoflagellate blooms can occur in sea ice, especially in the upper regions of fast-ice habitats in spring and early summer. At other times of the year dinoflagellates and chrysophytes are often found as stress-resistant (non-photosynthetic) cysts within sea ice. Diatoms can also form similarly resistant resting stages that can lie dormant in sediments for many years, although this is not common within sea-ice systems.

Apparently the cyst formation by dinoflagellates and chrysophytes is an adaptation for dispersal in the water column during summer and autumn, and also for overwintering in upper cold hypersaline ice. Dinoflagellates and chrysophytes within sea ice form cysts just prior to ice melt, and these are released into the surface waters where they are moved on surface water currents following the melt. Cysts are then incorporated into new formations of ice during autumn. Sometimes the cysts will develop into the photosynthetic stage in early autumn causing brief autumn blooms within the ice. These cells then re-encyst before the harsh conditions of winter set in.

Other photosynthetic organisms commonly found in sea ice include species of the colony-forming *Phaeocystis*. These are usually found in surface ice assemblages such as gap layers, melt ponds or in rotten summer sea ice. *Phaeocystis* occurs as individual photosynthetic cells embedded on the periphery of a water-filled gelatinous sphere. When these species bloom, either in the water or in the ice, thick soups of the spheres develop, which are not easily grazed by zooplankton.

Photosynthesis at low light levels

Photosynthetic organisms are able to grow in very low light conditions, down to water depths of around 260 m (853 ft), below which photosynthesis is impossible. However, on average, light penetrates to much shallower depths, generally between 50–100 m (164–328 ft). In sediment-laden coastal waters light may only be able to penetrate less than 10 m (33 ft). Ice is a most effective barrier to light, especially if it is covered by snow, and so sea-ice algae have to be physiologically geared to living at low light levels. Algae have been recorded as having the most extreme low light adaptations ever recorded. These algae flourish on the underside of ice over a metre thick covered by snow where the light reaching the algae in the bottom ice layer is less than 0.1% of the light at the surface of the snow.

At low light levels, photosynthetic organisms can increase their efficiency in utilising light by changing the amount and composition of pigments used to trap light energy. Algae absorb energy at 400–700 nm in the light spectrum called the 'photosynthetically available radiation'. All algae contain the green pigment chlorophyll a that absorbs in two parts of the spectrum: at wavelengths between 400–500 nm (blue to green) and 650–700 nm (red). There are many other pigments, called accessory pigments, often specific to individual classes of algae, that absorb light of different wavelengths than chlorophyll a, and these can be regulated to adapt to changing light quantity and quality. Sea-ice diatoms growing at low light increase the concentration of two accessory pigments: chlorophyll c and fucoxanthin. The latter is actually yellow-green in colour and is the pigment that gives diatoms their characteristic golden brown colouration. These pigments enable the cells to trap the wavelengths of light penetrating ice and snow (which chlorophyll a is less efficient in trapping). As light levels decrease, sea-ice algae can in addition increase the concentrations of chlorophyll a within the cells, thereby helping to ensure maximum absorption of the incident photons of light.

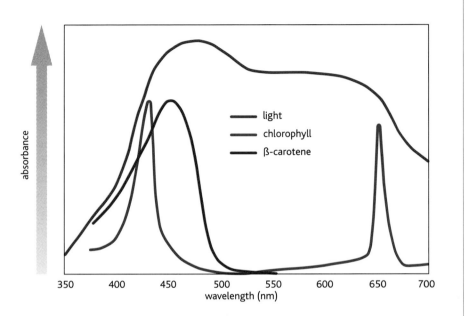

Top left
Wavelengths of light between 350 and 700 nm are used by algae for photosynthesis. Algal pigments such as chlorophyll and ß-carotene absorb wavelengths of light in different parts of the light spectrum.

Bottom left
Light within the ice limits how much photosynthesis, and therefore algal growth can take place.

Above
The quality of light is greatly altered as it passes through ice.

The spectral quality of light can be greatly altered as it passes downwards through a piece of ice. Violet and red portions of the light spectrum are absorbed preferentially by ice and snow, effectively narrowing the wavelengths of light that reaches deeper into the ice. The response of sea-ice diatoms to this spectral narrowing (which is usually accompanied by low light levels), is to increase absorption of green light by increasing photosynthetic accessory pigments (fucoxanthin and chlorophyll c).

Changing light levels and photoinhibition

In general ice algae have photosynthetic mechanisms that are saturated at low light levels and have high photosynthetic efficiencies. However, if the algae are then exposed to higher light conditions, at least for a period, damage to photosynthetic metabolism can take place, called photoinhibition. This is normally manifested in a decrease in the rate of photosynthesis, but if the transition from low light to high light is too extreme, or if the high light is associated with high levels of ultraviolet radiation (p. 97), the cells can be severely damaged.

Some accessory pigments such as β-carotene and diatoxanthin have a photoprotective role. These are used to protect cells from damaging high light levels and harmful ultraviolet radiation. Concentrations of these are usually low in light-limited algae, but are rapidly formed when the algae are transported into high light environments.

Another feature, exhibited by many algal cells when exposed to changing light conditions, is that the pigment-containing chloroplasts can move within the cells. The chloroplasts move along cytoplasmic strands in a process known as karyostrophy. Chloroplasts are usually distributed within the cell so that the most efficient light absorption can take place. However, clumping of chloroplasts, often around the cell nucleus, is frequently observed in high light intensities, and is thought to be associated with mechanisms to protect cell organelles from the damaging effects of light.

Surviving in darkness

During the extremes of winter the light levels in the polar regions are minimal, and darkness combined with low temperatures are the main features governing photosynthetic organisms during these times. Extended periods of total darkness may be experienced, especially if ice floes are covered by thick snow. Even when light returns in late winter/early spring, organisms within the ice may experience lengthened periods of darkness for several weeks, or even months in thick multiyear ice.

It is thought that in dark and/or very cold temperatures (and high salinities) organisms that cannot form stress-resistant spores or cysts enter into a resting stage. In a sense the process is somewhat like hibernation. During these phases cell metabolism is reduced to minimal levels that are just enough to keep the cells alive. In fact, some diatom species (from both polar and temperate regions) can survive periods of darkness for several months and even several years.

Below
In winter there is virtually no light available for photosynthesis to take place.

No photosynthesis is possible in the dark, although grazers are still able to feed on bacteria and algae and bacteria are able to divide and grow. As a result a significant amount of detritus (including faecal material, empty diatom frustules and breakdown products of grazing and bacterial activity) may accumulate in sea ice by the end of winter.

It has long been postulated that sea-ice algae actually switch from a photosynthetic mode to one where they take up dissolved organic matter such as glucose and amino acids as a means for winter survival. This has not been proven conclusively, although genetic engineering experiments have shown that diatoms have the necessary metabolic capabilities for utilising glucose and the potential to transform from obligatory photosynthesis to being able to grow without light using glucose as an energy source. Whether such switches by sea-ice algae are really possible is still open to much speculation.

Ice algae and nutrients

Photosynthetic organisms require a diverse range of elements besides carbon, hydrogen and oxygen for balanced growth. These include nitrogen, phosphorus, silicon, sulphur, potassium and sodium, which are all known as macro-nutrients. Many trace elements (micro-nutrients) are also required, including iron (p. 101), zinc, copper, manganese as well as vitamins such as for some species B12, biotin and thiamine.

Each nutrient, or rather the lack of it, has the potential to limit the growth of these organisms. In most aquatic environments it is either nitrogen or phosphorus that are generally the limiting elements, although growth can be limited by the supply of more than one nutrient at any one time.

Nutrients in the ice

There are several issues regarding nutrient uptake in sea ice that affect the growth of algal assemblages. The main issue is the rate of re-supply of a nutrient which is determined by whether or not there is the possibility of exchange with the surrounding seawater. In biological assemblages at the periphery of ice floes, especially bottom ice assemblages, nutrient depletion is seldom reported as being a growth-limiting factor, because there is an easy exchange with the underlying water. The uneven surfaces of the skeletal ice layer causes small-scale turbulence as the water passes under the ice promoting the exchange of nutrients and gases.

Bottom ice assemblages and Antarctic platelet layers support the highest standing stocks of sea-ice algae and this is largely due to the fact that the underlying water re-supplies the microbes with nutrients. Some diatoms, such as the Arctic species *Melosira arctica,* form large mats that hang down from the ice as strands up to 15 m (50 ft) long. These are often referred to as strand

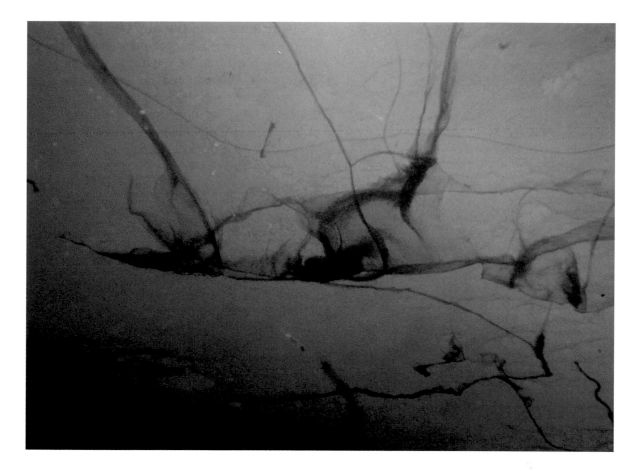

assemblages, and are not frequently reported in the Antarctic. This has to be the most efficient way to maximise nutrient uptake, using the ice as a physical support and then moving in the water. In effect, the assemblages are doing what a seaweed does.

Moving further up into the ice, exchange processes decrease proportionally with increase in distance from the ice-water interface. Even just 5–10 cm (2–4 in) from the ice-water interface there can be significant nutrient depletion and limitation of algal growth due to lack of water exchange. Therefore, interior ice assemblages, which are effectively cut off from this exchange, tend to use up any available nutrients and then their growth rapidly becomes inhibited.

However, ice floes are often broken and rafted together. Such large-scale physical transformations will at times expose these internal regions of the ice to nutrients in seawater. The deformations may also result in stress fractures and fissures within the ice, which act as conduits for nutrient-rich water to reach deep within ice floes, therefore supporting the growth of internal algal assemblages.

Above
Strands of diatoms hanging from the underside of Arctic ice floes and wafting in the passing water currents.

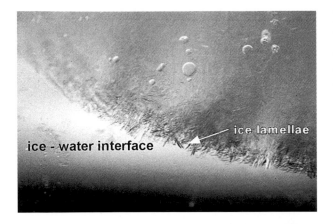

ice - water interface

ice lamellae

Top
The skeletal layer at the underside of the ice is a fragile structure built from loosely-packed ice crystals.

Above
Chemicals from the atmosphere can be deposited onto ice surfaces, and some can even be used to support algal growth.

Nutrients from the air

There is an input of nutrients from the atmosphere, since significant amounts are deposited and stored in snow. As snow is transformed and incorporated into the gross structure of ice floes, there is an input of the stored nutrients into the nutrient-depleted surfaces of the sea ice. The input of nutrients from the air will be greatest if there is industry nearby or a significant deposition of dust. Therefore nutrient input into the ice from this source is far greater in the Arctic compared to the Antarctic.

Gas production and consumption in a closed environment

During photosynthesis inorganic carbon (carbon dioxide) is taken up and oxygen produced. During respiration the opposite is true and carbon dioxide is produced and oxygen consumed. Algae and photosynthetic nanoflagellates and dinoflagellates all photosynthesise and respire, whereas bacteria and grazers (including copepods, amphipods and protozoa) only respire. The photosynthetic organisms can only photosynthesise in the light, but respire in both the light and in the dark. Because there is little chance of exchange of gases between the ice and surrounding water or the atmosphere above, a very different gas composition develops in these high biomass brines compared to normal seawater. They are characterised by substantial reductions in, and even exhaustion of, carbon dioxide, pH values up to 10, and very high oxygen concentrations. This composition indicates that photosynthetic activity is in excess of respiration by all of the organisms in the ice. However, to date measurements have only been made in spring and summer ice when this is likely to be true. In winter when light levels are low, at times of heavy grazing or mass mortality of algae with increased bacterial activity, this trend would decrease and eventually reverse.

Low concentrations of carbon dioxide can be experienced by phytoplankton in seawater, but supersaturated concentrations of oxygen (hyperoxic conditions) are very rare in marine systems. In hyperoxic situations harmful

Above
Measuring the high oxygen
concentrations in sea ice.

substances can accumulate, including chemicals such as hydrogen peroxide and hydroxyl radicals that can damage nucleic acids, proteins and other cell constituents. This is especially true when there are high levels of ultraviolet radiation present, which is the case for surface sea ice at certain times of the year. Sea-ice diatoms have been shown to have high activities of antioxidative enzymes that help to combat these potentially damaging conditions. DMSP and its breakdown products DMS and acrylic acid have also been shown to have antioxidative stress properties.

Sea-ice algae and ultraviolet radiation

Ozone depletion in both the Antarctic and Arctic are well documented, and the so-called ozone holes are responsible for increased levels of harmful ultraviolet radiation reaching the Earth. Ultraviolet radiation is normally split into UVA (wavelength 320–400 nm) and the more damaging UVB (wavelength 280–320 nm). UVB radiation can penetrate clear polar water to depths in excess of 50 m (164 ft). Although sea ice can effectively block UV light, experiments have shown that 1–2% of the incident UVB and over 5% of the incident UVA can reach the bottom of an ice floe 1.7 m (5½ ft) thick. However, when ice has even a modest covering of snow the amount of UV light passing through to the ice is very much reduced. Therefore, within the pack-ice regions of the world, a sea-ice cover is an effective filter, blocking most of the UV light and therefore protecting the organisms in the water below.

For sea-ice algae it is not simply the amount of UV radiation that is a problem. A high proportion of the algae within sea-ice are adapted to living in extremely low light levels making them highly susceptible to UV radiation damage.

Sea-ice algae and sunscreens

Algae are known to produce a number of protective UV-absorbing compounds including pigments such as ß-carotene and mycosporine-like amino acids (MAAs) in response to elevated levels of UV radiation. These act as sunscreens preventing UV damage to cell structures and DNA. Sea-ice diatoms and other ice algal groups have been shown to produce large amounts of these compounds, especially MAAs, when UV levels increase. Therefore sea-ice algae appear to be physiologically capable of dealing with the type of UV stress that may be found in ice.

Despite all of these factors UV radiation is not likely to influence sea-ice algae greatly due to snow cover and absorption by the ice. Some researchers have proposed that instead of causing major damage to sea-ice assemblages, UV radiation may alter the population structure so that species producing the greatest amounts of protective compounds dominate. It is also true that the highest levels of UV radiation do not necessarily coincide with maximum algal activity within the ice.

Overall estimates of productivity within the ice

The overall biomass of algae in water and sea ice is typically expressed as the concentration of chlorophyll a per unit area or volume of ice. Generally Antarctic sea ice supports higher algal biomass than Arctic ice. The highest algal biomass reported in Arctic sea ice is 300 mg chlorophyll a m^{-2}, although the average for all Arctic sea ice studies is 88 mg chlorophyll a m^{-2}. In contrast, concentrations in Antarctic sea ice have been measured in excess of 400 mg chlorophyll a m^{-2} and the average algal biomass in Antarctic ice is 133 mg chlorophyll a m^{-2}.

Antarctic platelet ice is the most productive sea-ice habitat, and maximum chlorophyll a concentrations in these habitats (>1000 mg chlorophyll a m^{-2}) is more than two times greater than that reported for any other ice type either in the Arctic or the Antarctic. Removing platelet ice from the average sea-ice algae biomass calculations for the Antarctic reduces the average to 88 mg chlorophyll a m^{-2}, the same as the estimate for the Arctic.

Primary production of polar oceans in a global context

One of the most basic measurements made by biologists working in aquatic systems is how much organic matter is produced by photosynthetic organisms. This is a fundamental measurement because it tells us about the total amount of biology that can be supported by a particular ecosystem. The amount of bacterial, algal or plant biomass (primary producers) built up over time through the process of photosynthesis is generally referred to as primary production. This is normally expressed as the amount of carbon fixed by photosynthesis, per unit area of space or volume, per unit of time. Most estimates are expressed as net primary production, which takes account of the respiration as well.

Global ocean net primary productivity estimates are numerous and varied. However, the most recent estimates tend to be around 40–60 Pg carbon per year (P = peta, and 1 Pg is equivalent to 10^{15}g). Of this, it is estimated that about 6.5 Pg of carbon are produced within the polar oceans each year. For comparison, recent estimates of terrestrial primary production are in the region of 50–60 Pg carbon per year. Combined with the oceanic primary production, this gives a total primary production for the Earth of around 100 Pg carbon per year. Therefore, the primary production of the polar oceans is a significant 6.5% of the total primary production on Earth.

The latest methods for estimating primary production mainly use satellite colour sensors to measure the concentration of chlorophyll a in the phytoplankton in the uppermost few metres of the ocean. These can then be converted to equivalent amounts of carbon. A major disadvantage of these techniques is that they cannot give information about concentrations of phytoplankton deeper in the water. In ice-covered waters the problems are even more pronounced, since the satellite technologies cannot measure the chlorophyll biomass inside the ice or in the water below the ice.

Estimates of large-scale primary production are difficult to make, and are based on *in situ* measurements of photosynthesis and respiration plus complicated mathematical models. The most recent estimates of total primary production in both Arctic and Antarctic sea-ice assemblages are very similar at around 70 Tg carbon per year (T = tera, and 1 Tg is equivalent to 10^{12}g).

Right
Annual production of algae in the Arctic Ocean (top) and the Southern Ocean (bottom). The highest production of algae is indicated by red and yellow, medium production by green and lowest by blue. The ice is white.

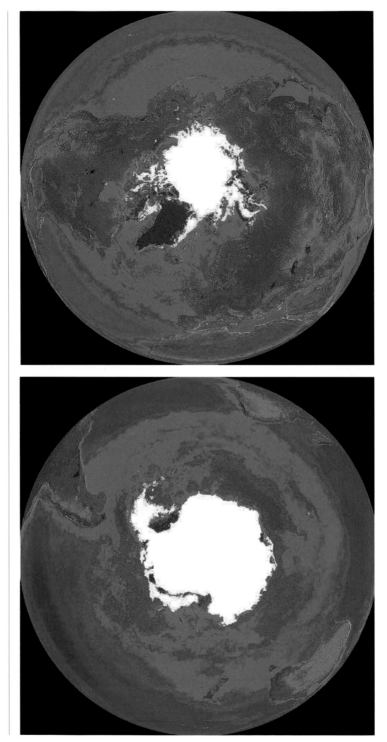

Using the Southern Ocean as an example, this is only a very small percentage of the total primary production, of the waters south of 50°S, which is estimated to be between 2900–4414 Tg carbon per year. In fact 88% of the annual primary production takes place in the permanently open ocean, the part of the Southern Ocean not impacted by sea ice at all. Although the ice primary production is only a small amount of overall Southern Ocean primary production, it is 10–28% of the total production in the seasonally ice-covered waters, estimates for which range from 141–383 Tg carbon per year.

Even though amounts of primary production by sea-ice algae are generally low compared to phytoplankton in open waters, perhaps it is not the amount that is so fundamental rather than its significance to the ecosystem. For long periods of the year sea-ice algae are the only food source for a diversity of protozoan and larger grazers. In fact, winter stocks of sea-ice algae have been shown to be vital for maintaining viable populations of many of these grazers, and some have life history and seasonal dynamics that ensure they can utilise this food resource when food in the water column is lacking. Therefore, in those waters that are covered in ice for part of the year, algae growing within sea ice are a vital component of the marine food web.

Iron-limited phytoplankton in the Southern Ocean

In some areas of the world's oceans the phytoplankton standing stocks are never large enough to deplete the nitrogen and phosphorus in the surface waters. These are the 'high-nutrient, low-chlorophyll' (HNLC) waters of the Southern Ocean, the sub-Arctic Pacific and the equatorial Pacific.

The low levels of phytoplankton seem strange given the abundance of nutrients, and several hypotheses have sought to explain HNLC. The most compelling explanation is that the rate of supply of iron, an essential trace element for phytoplankton growth, is limited in these regions. Dissolved iron concentrations in offshore areas are extremely low, since the primary source of iron to the surface waters of the oceans is from the land. The Arctic Ocean is not iron limited, due to iron inputs from the surrounding landmasses and large rivers discharging into the coastal waters.

Several oceanographic expeditions in the Southern Ocean have shown that during spring, phytoplankton bloom in iron-rich waters close to the polar front, but they do not bloom in waters with limited iron reserves. Three studies have now 'fertilised' Antarctic water bodies and all three found that diatoms bloomed in response to the added iron. This growth was in turn responsible for the absorption of significant quantities of carbon dioxide from the water during the experiments.

Ecological engineering and iron

It is the link between the phytoplankton growth and draw down of atmospheric carbon dioxide that fuels a vigorous debate about these experiments. There is a concern that these results may be viewed as providing a simple answer for mopping up excess carbon dioxide thereby curbing the effects of increasing greenhouse gases. By spreading iron over huge swathes of the ocean, enhanced phytoplankton growth would effectively trap carbon dioxide.

Such ideas about large-scale ecological engineering have little to do with the work of the scientists conducting the experiments. Iron fertilisation is in fact a poor way to tackle greenhouse gas problems, since levels of carbon dioxide are increasing at such a rate that even by maximising biological uptake in the oceans by adding iron there would still be net increases in atmospheric carbon dioxide.

The real interest of this work comes from the implications for the understanding of the atmospheric carbon dioxide levels in past climate history. This new evidence supports the theory that low amounts of atmospheric carbon dioxide (measured in ice cores taken in the Arctic and Antarctic) during past ice ages, may be linked to high amounts of iron in Antarctic waters that supported large standing crops of phytoplankton.

How about fertilising sea ice with iron?

There is an intriguing conundrum when interpreting the results of iron fertilisation experiments in ice-covered waters. If iron is limiting to the growth of phytoplankton in the Southern Ocean, and sea ice is formed from the same iron-deficient waters, it seems reasonable to conclude that the ice-based primary production should also be limited by iron? However, there is little evidence for this since sea-ice algal growth is rapid and reaches high biomass.

It is possible that along with all other brine constituents, dissolved iron is concentrated to high levels within the brine channels of sea ice. However, since a high proportion of brine is actually expelled from the ice, this is unlikely to supply enough iron to support the high algal biomass in the ice. Dissolved organic matter can form chemical complexes with dissolved iron which may make the iron more readily available to algae, and the high levels of DOM in sea ice (p. 106) would favour such processes.

Viruses and bacteria in sea ice

Viruses are now recognised as being important components in all aquatic systems, although it is only in the past decade or so that we have begun to understand their role in ecology and population dynamics within food webs. Viruses can infect a wide range of organisms from bacteria and algae through to larger organisms, and are now thought to affect species diversity and complex nutrient cycling within marine systems. In particular, viral infections of phytoplankton can have significant effects on the development and demise of seasonally important algal blooms.

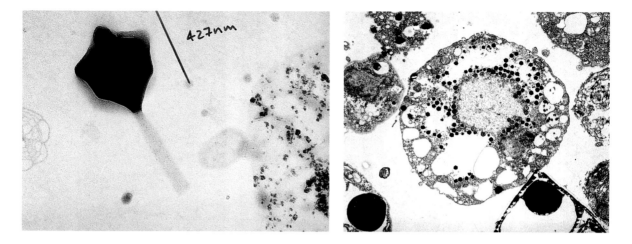

What are the hosts?

It is only since the 1990s that viruses have been described in sea ice. The few studies conducted to date have found that just like algae and bacteria there are higher numbers of free-living viruses in sea ice than in seawater. There are around 10^9 viruses per ml of brine, whereas seawater values are typically 10^6 per ml. The highest numbers of these viruses tend to be found where high concentrations of sea-ice algae are present.

Viruses obviously require host organisms for production, and free-living populations of viruses will gradually decline without the hosts. The most comprehensive virus studies from sea ice have measured large viruses (greater than 100 nm in diameter), which are thought to infect algae and protozoa within the ice. However, it is interesting to note that in complementary studies to determine which sea-ice organisms are actually infected with viruses, diatoms were not found to be infected. The only infected organisms were flagellate algae and other unidentified heterotrophic species.

Why aren't sea-ice diatoms infected by viruses?

In one study on Antarctic sea ice, no viruses were found to be infecting over 27,000 individual diatoms that were investigated. This would indicate that the diatoms are not actually a source of viruses within sea ice, which is somewhat counterintuitive considering that diatoms are the dominant component of sea-ice assemblages, and also because of the close correlation of virus numbers with ice-algae concentrations. Although the high numbers of viruses in sea ice indicates that they may play an important role in the food web dynamics, the low numbers of infected cells actually found would argue against this. It is thought that the high amounts of EPS (p. 88) produced by diatoms and other sea-ice organisms may be one of the reasons they are not susceptible to viral infection, the EPS providing an effective barrier to the viruses.

Above left
A virus isolated from Antarctic sea ice.

Above
A cell infected by many stained viruses.

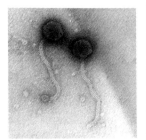

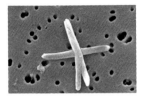

Above from top:
Electron microscope image of viruses that infect bacteria within sea ice.

Bacteria isolated from ice above Lake Vostok covered by Antarctic glacial ice. The sample was collected from ice 3593 m (11,788 ft) under the ice surface.

Diatom retrieved from 20,000 year-old ice from the Sajama ice cap in Bolivia.

The cyanobacteria *Aphanizomenon flos-aquae* from the Baltic Sea.

Bacteria infected by viruses

Bacteria are infected by viruses (bacteriophages), and the ratios of virus to bacteria (VBR) reported for sea ice are some of the highest seen in natural systems. Studies in the Arctic have shown downward shifts in these ratios with time, which indicates that as bacterial populations develop within the ice, virus-resistant species proliferate. Other studies have shown that sea-ice bacteriophages are host-specific, the phages only infecting one bacteria species. It is not just the host cells that are psychrophilic, but the phages as well. Therefore very specific host-phage interactions have developed within sea ice that are critically dependent on low temperature.

Bacteria thrive in sea ice

In contrast to the viruses, bacteria have been well studied in sea ice. The fact that high numbers of viable bacteria are found in sea ice is not surprising at all, since bacteria are known to inhabit some of the most hostile environments on the Earth. Bacteria are found thriving in thermal springs with temperatures reaching over 90°C (194°F), as well as in deep-sea environments at great pressures. Viable bacteria have been extracted from glacial ice over 20,000 years old from sites as diverse as Greenland, the Antarctic and high Tibetan plateaus. Most recently in the news are the bacteria that were retrieved from glacial ice 3.5 km (2 miles) deep in an Antarctic ice sheet overlying Lake Vostock. It is thought that the bacteria were isolated from the atmosphere many millions of years ago.

As a very general rule, bacteria are present in ocean waters at concentrations of about 10^6 cells per ml. Within sea ice, numbers up to 10^{10} per ml have been recorded. Therefore just as with viruses and algae, bacteria are present in the ice in concentrations very much greater than in the majority of ocean habitats.

Archaea and cyanobacteria

Archaea look like bacteria, but taxonomists have separated them into a separate kingdom from the true bacteria or eubacteria. Archaea have very different cell wall characteristics from bacteria, and their flagella with which they move around, are also completely different to those of the eubacteria. In fact it is thought that the archaea diverged from the eubacteria about three billion years ago. They are an interesting group of prokaryotic organisms, since they are typically found in hostile habitats such as boiling acidic waters, on and within crystals of salt, in deep-sea hydrothermal systems, and even within deep cracks in the Earth's crust. Therefore it is not unexpected that archaea are found in sea ice, and groups previously found only in hyperthermal environments (deep-sea hydrothermal vents and thermal springs) have been described living within sea-ice brines.

Another group of photosynthetic bacteria that are common in marine systems are the cyanobacteria (previously called the blue green algae). Just like the algae, these organisms can form blooms in surface water because they contain gas vacuoles in their cells, and can become a nuisance at times. For example, blooms of cyanobacteria are well recorded in the low salinity waters of the Baltic Sea. Typically, numbers of cyanobacteria are low in Antarctic marine waters and not often described in Antarctic sea ice (although they are important organisms in the lakes, ponds and glacial meltwaters of the Antarctic continent). Cyanobacteria are found in Arctic sea ice, although this is largely restricted to species colonising the freshwater melt ponds or living in association with ice where there is a large influx of freshwater such as the mouths of rivers. They are also found in sea-ice assemblages in the Baltic Sea.

Hitching a ride and gas production

It is thought that a high percentage of bacteria found within sea ice actually get into the ice by 'hitching a ride' on the surfaces of other organisms such as diatoms. Some ice bacteria exude carbohydrate glues to stick themselves onto diatoms, even puncturing the host's cell wall to ensure a good attachment. Another strategy is ice nucleation: the surfaces of bacteria themselves act as catalysts to promote the formation of ice crystals that then rise towards the surface and carry the bacteria into growing ice sheets. Some other sea-ice bacteria have been shown to produce gas vacuoles within their cells. Although the function of the vacuoles is still unclear it has been suggested that they may enable bacteria to rise up through the water column, bringing them into contact with sea-ice algae or even causing them to rise within the ice itself.

Anoxia in the ice and bacteria

The expulsion of oxygen during ice formation means that levels of oxygen can be low in sea ice (p. 25). In theory it follows that oxygen levels could be reduced to such an extent that conditions develop that favour anaerobic bacteria. When the respiration of the local sea-ice assemblage exceeds primary production, it is also thought to create anoxic microsites inside the brine channel system. Under these conditions all algae and normal ice bacteria and animals would be killed by the lack of oxygen. Only specialised bacteria that can live without oxygen would survive.

The evidence for this is still somewhat limited, although purple sulphur bacteria have been found inside ice cores from the Baltic Sea, and since these bacteria only grow in severely deficient oxygen conditions or where there is no oxygen at all, it seems reasonable to conclude that lack of oxygen may in fact play a role in determining ice bacteria populations. There are also several reports of hydrogen sulphide being smelt by researchers as they core the ice of the Antarctic. This again would indicate anoxic layers within the ice.

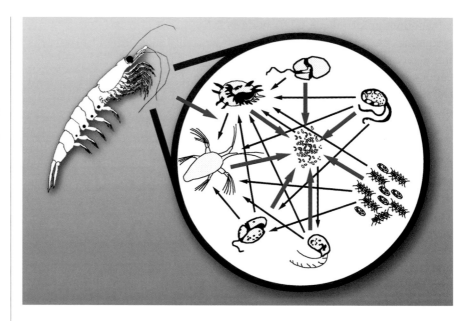

Bacteria and organic matter in the ice

The bacteria in marine systems, and also within the sea ice, grow by breaking down dissolved organic matter (DOM). The DOM arises from the excretion of organic matter by bacteria, algae and grazers, as well as the breakdown of cells when organisms die. Therefore the pool of DOM is a mixture of simple chemicals such as amino acids and glucose through to complex polysaccharides and cellulose cell wall material. Some of this material, such as amino acids and simple sugars, can be used immediately, whereas the more complex materials have to be broken down by enzymes released by the bacteria. Bacteria also need to assimilate inorganic nutrients such as nitrate and phosphate, but it is generally the supply of DOM that limits bacterial growth.

An important consequence of the bacterial breakdown of organic matter is the release of vital inorganic nutrients back into the surrounding water or, in the case of sea ice, the brines. These recycled, or remineralised, nutrients are then available to the algae to support new growth.

Development of sea-ice bacteria populations

Immediately after sea ice forms there is a general reduction in bacterial activity compared with rates in the open water. This is because the ice contains a mixed assemblage of bacterial species, many of which are not able to grow at low temperatures. With time the psychrophilic species (p. 73) begin to grow and bacterial activity rates increase and numbers increase. As an ice floe grows older and colder, bacterial activity generally continues to increase. Therefore, in young sea ice the ratio of photosynthetic carbon assimilated by algae to the

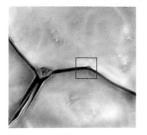

Left
Bacteria growing on the surface of a dinoflagellate from Antarctic sea ice.

carbon respired by bacteria tends to be high. In cold, older ice, especially in winter when photosynthesis is low, this ratio reverses and rates of bacterial respiration are generally greater than rates of photosynthesis.

Bacteria have been shown to be active in temperatures below –17°C (1.4°F) in Antarctic snow ice and down to –20°C (–4°F) in the Arctic tundra. However, it is now known that liquid phases exist down to –35°C (–31°F) in Arctic sea ice, and where there is liquid, it is quite likely that bacterial activity will be measurable, when we have the techniques with which to make such measurements.

Giant bacteria

Another general feature of sea-ice bacteria is that they tend to be much larger than bacteria from the open water. It may be that the large size is a low temperature response or that it is to do with the nutrient status within the ice. Most recent theories tend to say that the large bacteria are a result of low mortality rates within the ice. Many organisms from flagellates through to protozoa and crustaceans feed on bacteria, but bacteria within the ice can be somewhat protected from these grazers, enabling them to reach greater sizes.

Even within the ice many bacteria are not free living. Some studies have shown that up to 93% of the bacterial biomass in sea ice is actually attached to algae or other particles. Bacteria will also attach to EPS released by ice algae, thereby ensuring that they remain in close proximity to major producers of organic matter within the ice matrix.

Arctic vs Antarctic bacteria populations

With the advent of modern molecular techniques it is possible to scan field samples to obtain a good idea of the actual bacterial species composition. Comparison of bacteria species from Arctic and Antarctic ice indicates that not only do they both contain psychrophilic species, but there is a remarkable similarity in the species found at both poles. This naturally begs the question as to how the same cold-dwelling species have cropped up in both poles. Clearly transport across temperate and tropical waters is not a viable possibility. Transport in deep bottom waters is a plausible mechanism, although some of the species found to have a bipolar distribution are not tolerant of the high pressures found in the deep ocean.

There is a greater diversity of bacterial species in the Arctic sea ice compared to the Antarctic. This would appear to be a reflection of the greater variety of sources for bacteria in the former, including nearby land as well as the input of rivers. In fact the melt pond species on the surface of Arctic ice floes are completely different again from those found in sea-ice proper, and are most likely recruited from freshwater inocula via rivers.

Protozoa

Many different types of protozoa have been observed in sea ice, including amoebae, nanoflagellates, foraminifera, euglenoids, dinoflagellates and ciliates. They are found in high numbers in sea ice, many orders of magnitude greater than in the surrounding waters, where they feed on bacteria, algae and other protozoa within the rich biological assemblages. It is thought that the grazing by protozoa is highly significant in controlling the development and even species composition of algae and bacteria within the ice.

Bacteria guzzlers

Basically the food source of these grazing protists is proportional to the size of the grazer. Nanoflagellates range in size from 2–20 μm and move with the aid of flagella. They generally graze on bacteria, although they will ingest small algae. There are a few nanoflagellate species that can photosynthesize as well and these are called mixotrophic species.

The diversity of nanoflagellates smaller than 20 μm is vast, and marine scientists are still in the early stages of recognising and classifying the species. Understanding their ecology is at an even more basic stage. Choanoflagellates and small dinoflagellates have been shown to reach high biomasses in sea-ice, and these largely feed on bacteria as well as possibly taking up dissolved organic matter as a source of nutrients.

There is a large diversity of protozoa in the size range 20–200 μm, although the most studied are the ciliates and dinoflagellates. Ciliates have many forms, but in general they are spherical, oval or conical filter feeders with an array of cilia that are used for movement and wafting prey into the body cavity where they are digested. Some ciliates contain chloroplasts and are capable of photosynthesis and mixotrophy. One species, *Mesodinium rubrum*, which has been frequently reported in sea ice, contains chloroplasts and red accessory pigments and can only photosynthesise.

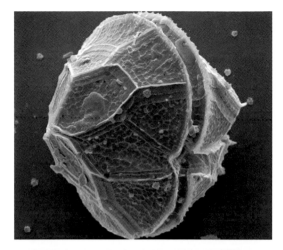

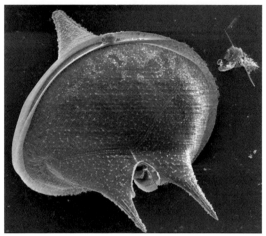

Dinoflagellates: armoured grazers

There are many dinoflagellates in marine waters, and many of these are photosynthetic species only, some others exhibit mixotrophy switching between photosynthesis and ingesting organic matter as an energy supply. However, there are a host of other species that live only by ingesting organic matter ranging in size from bacteria through to large diatoms and even chains of diatoms. They can also consume eggs and larval stages of copepods.

Above
Several dinoflagellate species are found in sea ice where they graze on bacteria and algae.

Dinoflagellates have two flagella and are strong swimmers. The largest can reach speeds of up to 1 m (3 ft) per hour. Typically dinoflagellates are oval- or pear-shaped although flattened species exist and others form chains of cells. Many species are covered in a series of armoured plates made of cellulose. These can mask the overall shape creating quite bizarre-looking organisms with an array of different types of horns and spines that make the term 'armoured' dinoflagellate quite appropriate. Other species do not have such cellulose plates and are referred to as 'naked' dinoflagellates.

In general sea-ice ciliates and dinoflagellates are restrained by the size of their food items. Dinoflagellates can ingest food items about the same size as themselves, whereas ciliates can only take in particles about 45% of their oral diameter, or about a tenth of their body volume. The consequences of this are highlighted by a study conducted within Arctic sea ice. The dominant grazing

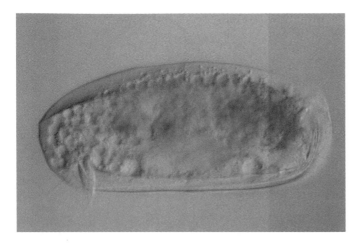

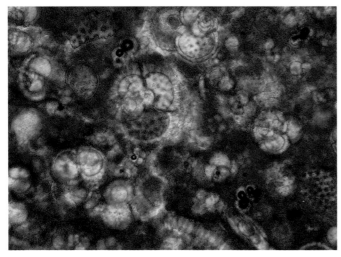

Top
The ciliate *Dysteria monostyla* isolated from Antarctic sea ice.

Above
The remains of foraminiferans sink to the seafloor where they collect to form sediments referred to as 'globigerina oozes'.

dinoflagellates were about 20–40 μm in diameter, and since the bulk of the sea-ice diatoms had lengths around 30 μm they were able to eat them. However, the main species of ciliate ranged in size from 30–60 μm in diameter (although the most numerous were between 30–40 μm). The ciliates were unable to ingest the diatoms, so presumably they were feeding on bacteria associated with the algal assemblages. Some Antarctic sea-ice ciliates have been shown to preferentially ingest bacteria. This is despite the fact that they could ingest much larger organisms.

Some open-water dinoflagellate species can ingest prey far larger than themselves. These produce specialised feeding tubes that they insert into the prey and use to suck out the cell contents. Others engulf the larger prey in a membrane, gradually digesting the prey within this veil. Whole chains of diatoms, far bigger than the dinoflagellate, can be digested at the same time using this method. Although these two methods of feeding have not been demonstrated within sea ice, it is most probable that some of the many dinoflagellate species found within the ice will use such methods of grazing, enabling them to feed on the larger sea-ice diatoms and even larval stages of copepods.

The flagellates, dinoflagellates and ciliates are all characterised by being highly motile. They are quite able to swim through brine channels, down to diameters that match their body sizes. Bacteria are not even safe in brine channels below 5 μm in diameter – if there are nanoflagellates present, the bacteria will be grazed upon.

Foraminiferans in sea ice

Another protozoan that is commonly found in Antarctic sea ice is the foraminifer, *Neogloboboquadrina pachyderma*. This organism is commonly found in pelagic systems around the world and occurs in such densities that in places the ocean sediments contain massive amounts of the undissolved calcareous outer cases (called tests) of the foraminifer that sink to the ocean floor. These accumulations can be so high that the sediments are often referred to as foraminiferal oozes that can cover massive areas of the seafloor.

Numbers of *N. pachyderma* can reach over 1000 individuals per litre of ice in the Antarctic. This is about 70 times the concentration of foraminiferans in the open water. Calculated in another way, 1 m³ (35 ft³) of sea ice can contain the same number of individuals as 60 m³ of the underlying water.

Neogloboquadrina pachyderma is the only species of foraminifer to be reported from sea ice, and one reason for this is its high tolerance to increased salinities. It is the only known species of planktonic foraminifer that can tolerate salinities up to 80. It is striking that although it is commonly reported in Antarctic sea ice in very high numbers, especially in bottom ice assemblages, it has been reported from Arctic sea ice in just a handful of studies. This is surprising considering that the species is found within Arctic waters. One of the reasons proposed for this anomaly is that *N. pachyderma* is not frequently found in waters of lowered salinity, such as those on the shallow Siberian shelf region of the Arctic where a high proportion of Arctic sea ice is first formed. Since studies have shown that it is caught up into the ice in the initial ice formation stages, rather than moving into the ice at later stages, this would at least partially explain the apparent paucity of the species in Arctic ice.

Left
The foraminifer *Neogloboquadrina pachyderma* is a common protozoan grazer in Antarctic sea ice.

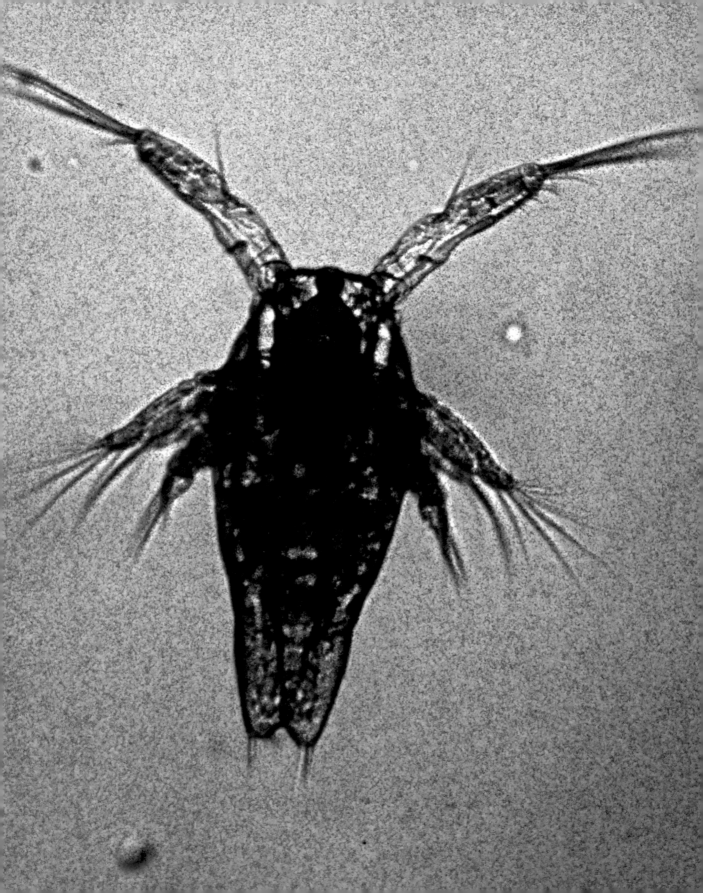

Animals in and around the ice

6

IT IS NOT JUST DINOFLAGELLATES, NANOFLAGELLATES, CILIATES, AND FORAMINIFERANS THAT GRAZE ON ALGAE AND BACTERIA IN SEA ICE. THERE ARE OTHER LARGER GRAZERS – THE CRUSTACEANS – REPRESENTED BY COPEPODS, AMPHIPODS AND EUPHAUSIIDS. ONLY COPEPODS ARE FOUND IN GREAT NUMBERS IN THE ICE; OTHER CRUSTACEANS MOSTLY FEED ON THE EDGE OF ICE FIELDS OR IN SURFACE MELT FEATURES AND GAP LAYERS AT THE TOP OF ICE FLOES.

COPEPODS CAN ACCUMULATE in concentrations of over 1000 individuals per litre of ice which, like the foraminifer concentrations, is considerably more than even the densest accumulations of copepods found in the open water (mostly less than 10 per litre). Copepods are restricted to parts of the ice with reasonably large brine channels and pores, since typical species found in sea ice vary in size from 0.5–2 mm long. This means that the highest copepod numbers are often found in relatively porous bottom ice assemblages, or in surface gap layers in summer perennial ice. Unlike turbellarian flatworms, copepods cannot alter their shape to squeeze into brine channels that are smaller than their body size.

Harpacticoid copepod species are commonly found in both Arctic and Antarctic sea ice. These species are typically found in surface sediments on the sea floor where they live in the interstitial water between sediment particles, or in the water layers just a few metres above the sediment surface. One species, *Dreschiella glacialis,* is found in extremely high numbers in Antarctic sea ice and can complete its whole life cycle within the ice where ice persists throughout the year. However, in the Antarctic this is restricted to only a few regions, and in other parts of the Southern Ocean pelagic or benthic life history stages of the species must exist. Adults are good swimmers although larvae and other young growth phases are not, and so how this species recolonises sea ice formed over deep water is unclear.

The copepods commonly found in sea ice are probably recruited via frazil ice. In fact some researchers think that certain copepod species, such as *Stephos*

Opposite
Juvenile stages (nauplii) of copepods and other crustaceans are commonly found in sea ice.

longipes, have especially sticky eggs that adhere to rising ice crystals, thereby ensuring that the species gets into the ice during winter. Other species of copepod have distinct ice and water components to their life cycles. *Stephos longipes* and *Paralabidocera antarctica* are the best studied Antarctic 'sea-ice' copepods, and it appears that their year long life cycles are closely synchronised to annual pack-ice dynamics. Even when sea ice persists throughout the summer these species do not complete their full life cycles without moving into the water for a period of time. However, both species maintain large numbers of young stages in winter ice that actively feed on sea-ice algae. The ice is therefore providing a resource to maintain a viable, actively growing population at a time when food supplies in the water column are scarce. When ice-edge blooms occur in spring and early summer these individuals are ideally situated to exploit the rich food supplies of the rapidly growing algal populations. The ice also provides some security from predators, thereby ensuring a greater survival rate of these young stages, which can develop into adults. The Arctic copepods *Tisbe furcata* and *Cyclopina schneideri* also overwinter in sea ice, although instead of spending summer in the water, they descend to live near the sea floor during the ice-free periods of the year.

Is a diet of sea-ice diatoms good?

It has been proposed – rather controversially – that a diet of diatoms may adversely affect the reproductive abilities of some copepods. This is thought to be because the diatoms produce certain aldehydes and these chemicals reduce the fecundity of copepod eggs. Clearly diatoms will form a large part of sea-ice copepods' diets, and it remains to be seen if sea-ice diatoms contain these aldehydes. If they do, sea ice will be an interesting system in which to test this proposed interaction, and to see if copepods that can complete their full life cycles in sea ice, such as *Dreschiella glacialis*, are actually immune to these effects.

Copepods not found in the ice

There are a large number of copepod species that, although not found in sea ice, evidently feed and remain just under the ice in the narrow band of water referred to as the ice-water interface. This is typically just a few metres thick, and is a dynamic region in terms of water chemistry. During freezing the ice-water interface is the zone of active ice growth, and brines (plus nutrients and DOM) from the ice sheet are expelled and diluted within this layer. In melt conditions these are the waters to be first stabilised by freshwater from the melting ice and, of course, are the first to experience the large fluxes of organic matter from breaking or melting ice. In particular, the skeletal ice layers are a rich source of ice algae and bacteria which are available to copepods during ice-covered seasons when food in the underlying water is scarce.

Boundary layers

In aquatic systems, water passing close to solid objects effectively slows down due to the friction and properties of the water. At the surface of the object the velocity of the liquid is effectively zero. Moving away from the surface the velocity of the liquid increases to the free stream value. The region of reduced water velocity is called the boundary layer because it occurs at the boundary of the fluid and the solid structure. The boundary layers at the ice-water interface may be important refuges for organisms.

Species that are not found in sea ice, but that do exist in the ice-water interface, may not move into the ice because of their larger size, up to 1 cm, although this is not exclusively true. *Oncea curvata* and *Oithona similis* are small copepods approximately 1 mm long. They are seldom found in sea ice, but frequently in the ice-water interface. Size clearly isn't the problem for these species, but possibly they are not able to cope physiologically with the salinity and temperature changes that take place even in the skeletal layer.

Two larger Antarctic copepods, *Metridia gerlachi* and *Calunus propinquus,* are frequently found in the ice-water interface, but never in sea ice. Even a very slight increase in salinity has been shown to be lethal for these two species, and this is why they are restricted from moving into the ice. However, they have been shown to feed on ice diatoms as they are released from the ice. It is possible that this will account for why *Pseudocalanus* species, *Calanus glacialis* and *Calanus hyperboreus* copepods are found in high concentrations under Arctic ice, but never within the ice matrix.

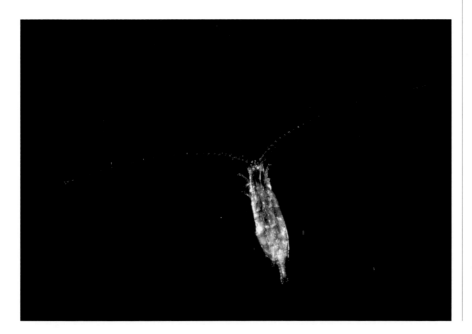

Left
A copepod *Rhincalanus giga* that is too big to go into the ice, but can be found underneath ice floes.

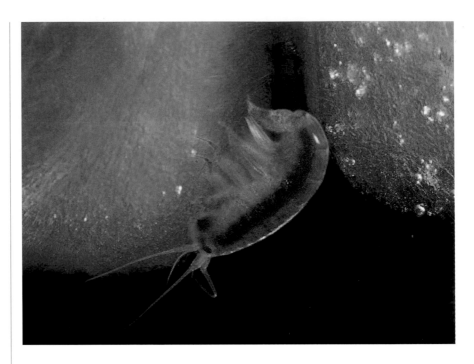

Amphipods

A variety of amphipods are recorded as actively feeding on ice algae on the underside of ice floes in both the Arctic and Antarctic. Most of the species found connected with sea ice are herbivores. However, other species eat a mixture of algae and animal remains. The Antarctic *Paramoera walkeri* and the Arctic *Gammarus wilkitzkii,* which are known to eat ice algae, have been observed predating on copepods and amphipods – even individuals from the same species.

The Arctic *Gammarus wilkitzkii* and *Apherusa glacialis* in particular reach very high numbers under ice floes. It would appear that these two species require summer sea ice in which to complete their life cycles, whereas most other amphipod species associated with sea ice are not dependent on sea ice, but rather use it as a feeding ground and as a refuge from predators.

In contrast to the Antarctic, euphausiid species are not found in huge swarms in the Arctic, and are not found grazing on ice algae. Instead amphipods are the most dominant under-ice grazers in the Arctic, reaching concentrations equal to the densities of krill found underneath Antarctic ice floes.

Such amphipod species are also recorded on the underside of Antarctic ice floes, but generally only those overlying shallow coastal regions. In particular *Orchomene plebs* can reach high numbers, especially in fast-ice sites and underlying layers of platelet ice. Most of these species are normally found on the benthos, but being good swimmers they are able to swim up to the ice and use the same modes of feeding as they do on the sea floor, simply upside down. In the Antarctic *Paramoera walkeri* and *Pontogeneia antarctica* migrate from the sea floor when sea-ice forms, and their young are released onto the underside of the floes. The young feed under the ice over the winter, thereby enhancing the chances of survival in the next generation.

Again, because of their large size (up to several centimetres long), amphipods are rarely found in sea ice itself, although very high numbers have been recorded from fairly solid platelet layers in the Antarctic. The Arctic *Gammarus wilkitzkii* and *Onismus glacialis* are killed by being frozen into sea ice, even for short periods. However, *Gammarus oceanicus* can actually tolerate being frozen in ice.

Fat reserves

All crustaceans store energy reserves in the form of lipids. The Antarctic krill can store up to 44% of its biomass as lipid, generally utilised when food in the water is low. Species of amphipod and copepod that stop feeding during winter store mainly a class of lipid called wax esters. These are typically used for long-term energy supply and are usually found in species that cease feeding for extended periods, such as in winter months when food is scarce in the water column.

In contrast those species that continue to feed during winter use another class of lipid, the triacylglycerols. Most of the species of copepod and amphipod feeding directly on ice algae from within or on the underside of ice floes use triacylglycerol. Even those copepods that do not move into the ice, but are only found in the ice-water interface have been shown to use triacylglycerols. This indicates that the supply of ice algae in the shallow water layer during winter is supporting active metabolism of these pelagic copepods, when other wax-storing species deeper in the water column are living off their lipid reserves.

Above
Copepods such as *Calanus propinquus* lay down fat reserves to survive periods when food in the water is limited.

These grazers are not able to produce polyunsaturated fatty acids (PUFAs). As described before (p. 74) ice diatoms and bacteria are a rich source of PUFAs and so the crustaceans feeding on the ice algae in winter not only have a rich source of food, but one enriched with a class of lipid that they need for maintaining membrane integrity and functioning at low temperatures.

Sea ice and krill

There are several species of euphausiid crustaceans in polar waters. One of them, *Euphausia superba* (or krill) is present in huge stocks in the Southern Ocean, estimated to exceed 1.5 billion tonnes (the total mass of all the people on the Earth is approximately 0.5 billion tonnes). Krill are the primary food for squid, penguins, some seals and baleen whales. In fact they are often referred to as being a key species in the Southern Ocean food web. Krill feed voraciously on phytoplankton and adults can also feed carnivorously on copepods. Krill

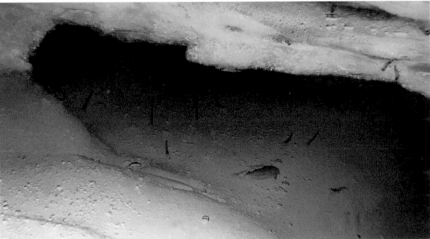

Above right
Euphausia superba, the Antarctic krill.

Right
Krill feeding on the underside of an ice floe.

populations can form dense swarms, so that if they come to the surface they can turn the water a spectacular blood red colour. A comparison between a krill swarm and a swarm of locusts does not seem unreasonable. Both swarm in huge numbers and can strip the food effectively from every area they travel through.

Starvation or a diet of sea-ice algae

Adult krill can survive extended periods of starvation, living from lipid reserves and even shrinking over the winter. Although adult krill can be found in association with sea ice, the most striking link between sea ice and the species is the distribution of larvae and juvenile stages. These stages do not have enough reserves to starve for any considerable time. The sea-ice algae and bacteria assemblages on the underside of ice floes are a vital food reserve for these creatures, enabling them to survive the winter months when food in the water is absent. Krill are voracious feeders, scraping the ice algae from the ice surfaces with great efficiency. In fact larval krill can ingest up to 44% of their body carbon content per day, thereby ensuring that when the ice melts in spring and food is available in the water once again, they are in good physiological shape to grow at maximum rates through the summer.

Below
Krill are often termed the central or 'keystone' organism in the Southern Ocean food web.

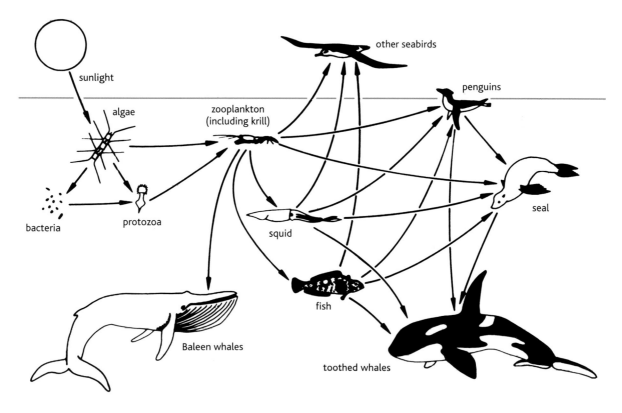

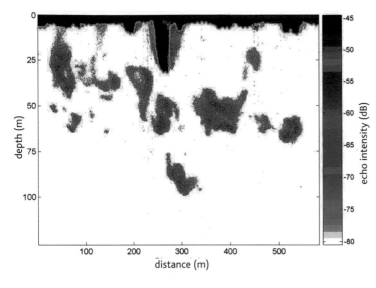

Above
600-m (1969-ft) transect under ice floes by *Autosub* equipped with an echosounder that showed huge swarms of krill lying mainly about 50 m (164 ft) below the ice. A 25-m (82-ft) deep pressure ridge keel shows up on the under-ice profile.

Using submarines to look for krill

One of the problems with studying krill is measuring the distribution of the krill under the ice. Recent developments in AUV technology (p. 84) mean that it is now possible to deploy them reliably under pack ice to conduct extensive echosounder-surveys covering areas of many hundreds of square kilometres. Krill swarms are easily detected using echosounders looking downwards, while upward-looking sensors detect the thickness of ice above and its concentration of krill. In a study in the Weddell Sea in January 2001 these techniques showed that the density of krill was three times higher under the ice compared to the adjacent open water with most of the krill within a band 1–13 km ($^2/_3$ mile to 8 miles) south of the ice edge.

Sea ice is not just a source of food for krill populations, but the ice also provides a refuge from predators. High densities of krill have often been observed to congregate in the regions around the keels of pressure ridges, and under deformed and rafted ice floes. The krill can hide away in the cracks and crevices, effectively hidden from feeding penguins, seals or fish. Typically flat, undeformed sea ice will have very few, if any, krill on its underside.

In disintegrating summer sea ice, krill are able to swim through the rotten layers of ice into the highly productive subsurface gap layers, and even into melt ponds. In the latter they are moving from seawater salinities of 34 to very low salinity water (below 10) with no apparent acclimatisation period. This example demonstrates that krill must have very efficient physiologies to cope with rapid changes in salinity, a feature not commonly displayed in ocean-dwelling crustaceans. Again by feeding in these layers the krill are not only exploiting a rich food source, but they are also gaining refuge from predators.

Good and bad krill years

Krill are central to the viability of the many animals and birds that feed on them. For this reason, the distribution of krill in Antarctic waters has been studied for several decades. It is now clear that the distribution patterns of krill are closely linked to sea-ice conditions. In years when sea-ice cover is prolonged there is significantly higher krill recruitment, and in some regions of the Antarctic, such as the Antarctic Peninsula region, the abundance of krill can be predicted on the basis of cyclical variations in sea-ice extent.

Salps are floating marine tunicates with a transparent gelatinous body that has an opening at each end. The salp, *Salpa thompsonii,* is thought to reach high densities in years following reduced ice extent. Salps live for less than one year, as filter-feeders of phytoplankton. They do not feed on ice organisms. In the absence of krill, the salps are able to exploit the spring phytoplankton bloom and undergo explosive population growth. In good sea-ice years, the krill have the upper hand over the salps because the sea ice has provided good feeding grounds over the winter, resulting in good gonad development and possibly allowing multiple spawning to take place. In these years the krill exploit the phytoplankton bloom resulting in poor food stocks for the salp populations.

Reduced ice and krill stocks

It would appear that krill stocks have decreased significantly around the Antarctic Peninsula region since the 1960s. However, the explanations for this are unclear. There have been reductions in sea-ice cover that persists through the summer in the Bellingshausen Sea over the same period, but it still remains to be proven that the decreases are real, or simply reflections of cyclical trends in populations possibly tied in with long-term cyclical sea-ice dynamics, such as the Antarctic Circumpolar Wave (p. 60).

It has been suggested that krill stocks are not actually related to ice cover *per se*, but rather linked with the total length of the ice edge. For instance, it has been estimated that if there was an overall 25% decrease of sea ice in the Southern Ocean, this would only result in a 9% decrease in the length of the ice edge, and therefore may not have such a dramatic effect on krill distributions.

It is not just sea ice that will have large-scale influences on krill populations. Krill swarms are reported to sometimes be washed up on shores or undergo

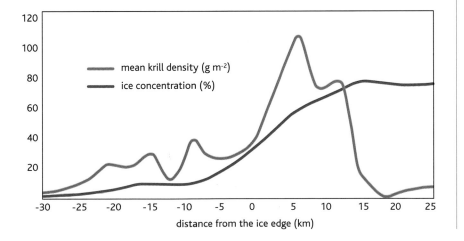

Left
Graph to show that the maximum density of krill occurs from the ice edge up to 10 km into the pack ice.

mass mortality events where whole swarms die. One of the explanations for these mass deaths is that krill are susceptible to infection by ciliate parasites, which can kill the host within 40 hours. The parasites rapidly divide within the host and then the animal bursts open to release swarms of infective stages.

The swarming behaviour of krill increases the chances of transmission of the parasite and other diseases.

Krill and sunscreens

It is also thought that krill may be more susceptible to damage from harmful UV radiation compared to other zooplankton species in the upper 20 m (66 ft) of the water column. Krill do migrate in the water column during the day: in daylight they are found at greater depths, swimming towards the surface at night. When there is a lot of UV radiation there is evidence that krill may move deeper in the water during day, thereby avoiding the harmful UV radiation.

It has already been described how sea-ice algae can produce pigments and compounds such as mycosporine-like amino acids (MAAs) in response to high UV radiation levels (p. 97). By grazing on sea-ice algae, krill and other zooplankton are able to incorporate MAAs into their diet and therefore incorporate the sun-screen benefits into their bodies. It has been estimated that krill can obtain at least 10 times more MAAs per unit of biomass ingested when consuming sea-ice algae than they can from eating Antarctic phytoplankton. The ice algal MAA source may be particularly important to krill during the austral spring when ozone levels are still low, sea-ice algae are actively growing, and the major phytoplankton blooms of the Southern Ocean have yet to develop. Because MAAs can be transferred from grazers to their predators such as seals, whales and penguins, the protective compounds produced by ice algae have the potential to benefit organisms over a whole range of trophic levels.

Nematodes and turbellarian worms

For predominantly benthic organisms such as turbellarian flatworms, harpacticoid copepods and nematode worms, the mechanisms by which they are incorporated into sea ice are not always clear. The organisms that live between surface sediment grains on the sea floor are generally poor swimmers and many do not have an obvious life in the open water. One possible route is that residual multiyear ice contains populations of organisms that act as inocula for newly-formed ice. This theory is fine in the Arctic where there is a great deal of multiyear ice, but in the Antarctic only a very small percentage of the sea ice lasts for more than one season, so this seems unlikely.

In coastal regions with shallow water depths it is not difficult to imagine colonisation of the sea ice from the benthos by larval stages, and even species with poor swimming capabilities. Another commonly cited vector in shallow water is the lifting of organisms from the benthos attached to anchor ice formed in shallow coastal waters. In fact some large benthic invertebrates such as starfish and polychaete worms are incorporated into sea ice in this way. However, most of the pack ice, especially in the Antarctic, overlies water several thousand metres deep and here mechanisms of colonisation by non-planktonic organisms remain enigmatic.

Flatworms

Turbellarian flatworms are found in both Antarctic and Arctic sea ice, although they are far more prevalent in the former. In a synopsis of many Antarctic studies turbellarians were found to account for over 50% of the total carbon biomass of the grazers in diverse types of sea ice, even though in actual numbers they make up about 25% of grazers in the ice. In contrast, nematode worms form a very important part of Arctic sea-ice assemblages, but have only been recorded in Antarctic sea ice on one occasion. In land-fast ice in the Arctic, nematodes can

Below
Flatworms can migrate through ice, altering their body sizes in response to the salinity of the brine.

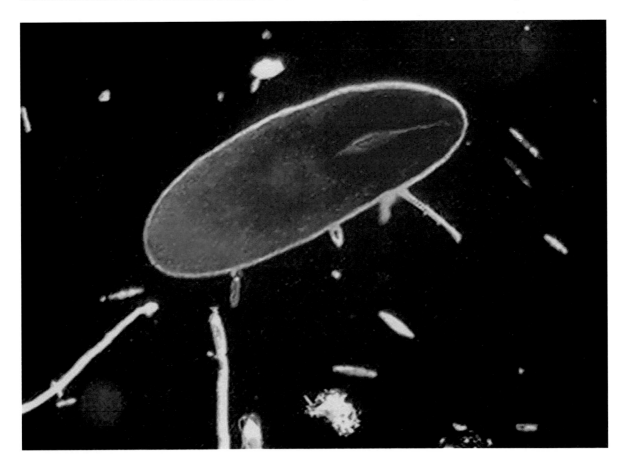

reach very high numbers indeed, dominating the community of small grazers, and accounting for up to 96% of all the grazers in some bottom-ice assemblages.

All of the nematodes described from sea ice come from the group Monhysteroida. These vary greatly in size, but species isolated from the ice are generally 1–3 mm long. Other members of this group typically inhabit marine and freshwater sediments, decaying seaweed and relatively dry soils. Some of them are tolerant of changing external salinity conditions. Interestingly the most conspicuous nematode species in ice have never been found in sediments. Generally they are poor swimmers, and in anything but the shallowest of sites it is improbable that nematodes swim from sediments to overlying sea ice. The fur of mammals, plumage of seabirds and even whale baleen plates contain thriving populations of nematodes (including Monhysteroida species) and these have been cited as possible vectors for bringing nematodes into close contact with sea-ice floes.

Species of nematodes isolated from sea ice can feed on sea-ice algae, bacteria and protozoa, while some of the larger species may feed on smaller nematodes. However, some of the species have never been found with anything in their guts and it is thought that these individuals may not be feeding directly on other ice organisms, but instead they are absorbing dissolved organic matter as their source of nutrition. The rich DOM concentrations within the ice would certainly support this hypothesis.

Moving within the ice

Turbellarians vary greatly in size, but species isolated from the ice generally are 1–3 mm long and less than 0.5 mm in diameter. This shape makes them ideal for traversing through skeletal layers of sea ice and into and through relatively small brine channels.

Sea-ice turbellarians have a very wide salinity tolerance, certainly surviving salinities up to three times seawater values. As they move into more saline waters they regulate their body tissue water content to compensate for the changes in salinity. This causes their bodies to shrink in proportion to the salinity of the water, resulting in the useful ability to traverse spaces and brine channels significantly smaller than their usual body size. This means that turbellarians have access to large regions of the ice for grazing on ice algae and bacteria. However, it has been estimated that brine channels 0.2 mm in diameter or less are effective barriers to these grazing organisms. The bacteria, algae and protozoa living in such channels will therefore be growing in an effective refuge from grazing pressure.

Another aspect that may prevent the grazers effectively feeding on the ice assemblages is the accumulations of EPS surrounding ice diatoms and bacteria (p. 88). These viscous substances may restrict movement, or simply deter the grazers from trying to feed.

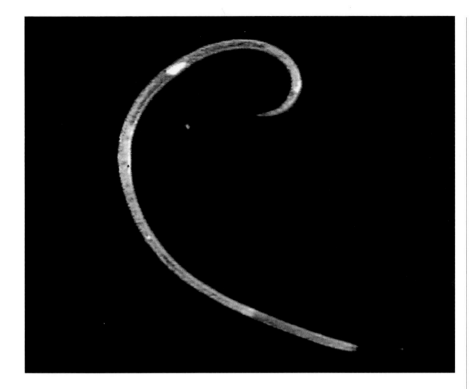

Hitching a ride on the back of a crustacean

Antarctic turbellarians spawn in sea ice in austral summer. Eggs, juveniles and adults will be released into the water column when the ice melts. Although sea-ice turbellarian species can swim, they have never been reported in plankton and it is presumed that they sink to the sea floor when sea-ice breaks up. It has been suggested that sea-ice turbellarians may have an adhesive disk by which they attach themselves to crustaceans such as amphipods or copepods before being released from the ice. Swimming crustaceans, including amphipods that migrate from the sea floor to the ice peripheries, or the common ice copepods may act as vectors to transfer the flat worms into newly formed sea ice and between ice floes in developed pack-ice fields.

Why are there so few nematodes in Antarctic pack ice?

The fact that nematodes are not commonly found in Antarctic sea ice in contrast to the high numbers in the Arctic is intriguing. No rotifers have been found in Antarctic sea ice either, even though these too are also common in Arctic sea-ice samples. It is possible that these Arctic/Antarctic differences are simply sampling artifacts, and in time more comprehensive sampling may produce more complete faunal records for Antarctic sea ice. Foraminiferans, which are very abundant in Antarctic sea ice, had for many years remained unknown from

Arctic sea ice in nearly 100 years of study. Since the late 1990s, however, foraminiferans have been found in Arctic pack ice, albeit only in a few samples. Alternatively, it may be that in the Antarctic there are no suitable vectors for the colonisation of sea ice by rotifers and nematodes.

Fish

Several species of fish are found close to ice floes. Most notable in the Arctic is the arctic cod (*Boreogadus saida*) and the glacial cod (*Arctogadus glacialis*). In the Antarctic the broadhead fish (*Pagothenia borchgrevinki*) feeds on the underside of sea ice and in platelet ice. These fish have been observed clinging to the underside of the ice when resting, and use crevices and holes in the ice as hiding places. Young stages of the Antarctic silverfish (*Pleuragramma antarcticum)* have also been shown to feed on the underside of ice floes as have, on few occasions, young stages of the giant Antarctic toothfish (*Dissostichus mawsoni*). These two species normally live in deeper waters, but migrate to the underside of the ice to feed on zooplankton.

Generally it is the younger fish that are found closely associated with the ice, since the principal food items in the ice-water interface are copepods,

Below
The Antarctic silverfish, *Pleuragramma antarcticum,* has been shown to feed under ice floes.

euphausiids and amphipods. Several studies have investigated the gut contents of the small fish *Pagothenia borchgrevinki* living close to ice; copepods and amphipods were the main food items, although larvae of other fish (*Pleuragramma antarcticum*), euphausiids and pteropods were also found.

The importance of this ice-related food for juvenile fish stocks is clear, since *Boreogadus saida* is a key fish species in the high Arctic, and in the Antarctic species such as *Pleuragramma antarcticum* are a vital food source for many birds and seals. The latter has a circumpolar distribution, and it dominates the pelagic fish species, accounting for over 90% of the fish biomass in the Ross Sea.

Many of the other Antarctic pelagic fish species have been shown to feed extensively on krill. Therefore the importance of sea ice for maintaining krill stocks has an important consequence for these krill-eating species of fish.

Fish antifreeze

The freezing point of the blood of many fish is about 1°C (1.8°F) higher than the freezing point of the surrounding seawater. This means that without modification fish blood would freeze at around –0.8°C (30°F) whereas seawater freezes at –1.86°C (28.65°F). Most polar fish are able to produce antifreeze proteins or glycoproteins which lower the freezing point of the blood to values close to the freezing point of seawater. There is a wide variety of these chemicals and the broadhead fish, *Pagothenia borchgrevinki*, alone produces eight different types of glycoproteins. There is evidence that Arctic fish generally synthesise their antifreeze compounds in winter, whereas the Antarctic fish produce them all year round. However, those individuals living in close association with sea ice have the highest concentrations of antifreeze molecules in their blood, independent of the season.

The common name of 'ice fish' would suggest that Channichthyidae actually live in the ice. This is not the case, but these ghostly white fish have unique physiologies for living at cold temperatures. They do not have haemoglobin-containing corpuscles in their blood, but instead oxygen is carried in blood plasma in large blood vessels. This is possible because at the low temperatures oxygen saturation of the blood is greater than in warmer waters. The ice fish also have a much greater volume of blood per unit of body mass and larger hearts than other fish species ensuring adequate oxygenation of body tissues.

Above
The Antarctic mackeral icefish, *Champsocephalus gunnari*, does not have haemoglobin-containing corpuscles in its blood, but instead oxygen is carried in blood plasma in large blood vessels.

Are there unique ice organisms?

Despite the unique physiological and biochemical prerequisites for living in the low temperatures, low light and high salinities found within the ice matrix, it must always be remembered that the sea-ice biology is recruited from the open water. As such it is peculiar to think of endemic species that are restricted to the ice. On the other hand, it is not too difficult to imagine that over evolutionary timescales species may have evolved that are restricted to living within the ice.

Clearly there are some organisms that are better suited to living within the ice than others, but there has been a surprising lack of evidence of species of bacteria, algae or protozoa that are found only within the ice. One exception is

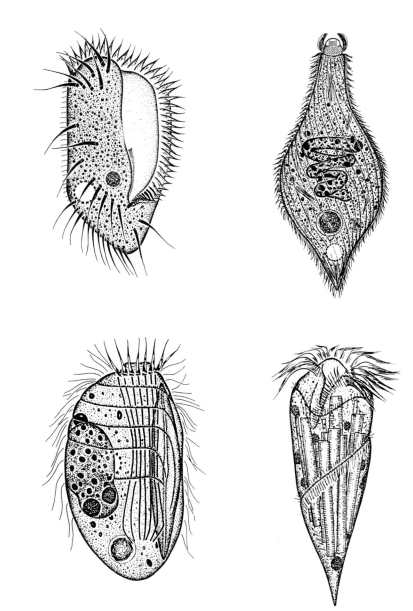

Right
Many ciliate species such as *Cytharoides balechi* and *Lacrymaria spiralis* are often found in large numbers within ice floes but rarely, if at all, in the surrounding water.

the Antarctic ciliates. Many ciliate species have been described from sea ice that have never been reported in the plankton. In one comprehensive study 68 species of ciliate were described; 55 of these were found only within the ice, six were restricted to the water and seven species were found in both water and ice. It is speculated that for many of these species colonisation of the ice takes place via resting spores that float in water and only the adult phase exists in the ice, although there is no direct evidence for this.

To a large extent the lack of evidence for endemic ice organisms may simply reflect the lack of studies looking for such species, and with the advent of modern molecular techniques it is likely that many new endemic microbial species will be described. This will be especially true for the bacteria, as more bacterial species are isolated from sea-ice habitats and described. Clearly many psychrophilic bacteria species must exist in surface waters either as dormant stages or as cysts. It is only when they are incorporated into the freezing ice matrix that they flourish.

Are they there, or is it that we simply have not looked?

An example of the type of discoveries that probably await us was the finding in the 1990s of a new dinoflagellate species. *Polarella glaciali* was isolated from sea ice in McMurdo Sound in the Antarctic. The species' cysts are remarkably similar to ancient fossil Suessiaceae cysts that date back to the Triassic and Jurassic periods. To this day *P. glacialis* is the only living member of the Suessiaceae. However, in 2003 further work identified a strain of this species that had been isolated from Canadian Arctic waters. So now we have a species found in Arctic waters, Antarctic sea ice and ancient sediment samples. This is typical of the conundrums facing researchers trying to understand biogeographic distributions and evolution of marine organisms. Part of the problem is due to the fact that the numbers of samples we have looked at are relatively few, and much of the time we simply do not know what is there because we have not looked.

Life under the ice

7

THE SEASONAL CHANGES IN SEA-ICE DISTRIBUTION AND
THICKNESS ARE NOT JUST IMPORTANT FOR THE SEASONAL
DYNAMICS OF ORGANISMS CAUGHT UP IN THE ICE MATRIX.
WHEN SEA ICE IS PRESENT IT FORMS AN EFFECTIVE BARRIER
BETWEEN THE ATMOSPHERE AND UNDERLYING WATERS, WITH
CONSEQUENCES FOR THE ORGANISMS THAT LIVE IN THE WATER,
AND ON AND WITHIN THE UNDERLYING SEAFLOOR SEDIMENTS.

ULTIMATELY THE BIOLOGY contained within the ice is released into the water when ice melts, and this either continues to exist in the water, or sinks to the sea floor. When large regions of ice melt as they move into warmer waters, the freshwater released lowers the salinity of the surface water, which effectively floats on the surface of the ocean. Plankton may become trapped within this stable layer of low-salinity water and instead of being mixed by water movements through the top 100–200 m (328–656 ft), they are suspended in shallow waters of below 50 m (164 ft). For photosynthetic algae, this means that they encounter higher light levels than normal, which induces high rates of photosynthesis and growth.

These higher rates of growth can lead to the so-called ice-edge blooms that are characteristic of melting ice edges. These are frequently seeded by diatom species that are being released from the ice, and there are many instances where the dominant algal species within the ice is the species that blooms in the adjacent stabilised water layers.

Such ice-edge features are not always guaranteed, because even though the surface waters are stable they can be easily broken down by strong winds and waves. For the algal blooms to take place the surface water layers need to be stable for periods of several days to a week, since most of the common diatom species generally double their population size at rates of just over once a day at the temperatures that would be encountered in such waters.

When ice-edge blooms do occur they characteristically attract zooplankton, such as copepods and krill that feed on the rich phytoplankton food source. In

Opposite
The giant Antarctic isopod,
Glyptonotus antarcticus
grows up to 20 cm (8 in) long
and 70 g (2½ oz) in weight.

turn, predators of the zooplankton, including fish, seals and birds, will home in on the plankton-laden waters to enjoy a concentrated food supply.

Sea ice, phytoplankton stocks, pelagic molluscs and an iceberg?

In both boreal and austral late spring and summer the growth of phytoplankton in the surface waters is a major component of the overall productivity of the Arctic and Southern Oceans (p. 98). This production is vital for supporting the food web, and is intrinsically linked with the seasonal dynamics of the sea-ice cover.

This is highlighted by a study on phytoplankton availability and the stocks of pelagic marine molluscs, known as pteropods, which feed on phytoplankton after trapping them in huge webs of mucus. In the austral summer of 2000/2001 there was a reported 50–70% decline in the overall phytoplankton biomass of the western Ross Sea, linked to the slowing down of sea-ice decay due to the break off of the large iceberg B-15. The reduced phytoplankton stocks led to decreased activity and reproduction in the pteropod *Limacina helicina,* which was not recorded in McMurdo Sound for the first time since records started. In turn, populations of another pteropod, *Clione antarctica* that feeds directly on *L. helicina,* were greatly reduced. These pteropods are important as part of the diets of fish and whales in the region. This illustrates the cascading impact on the food web due to the altered phytoplankton populations brought about by sea-ice dynamics.

Below
The remains of organisms sink rapidly through the water in aggregates of material often called 'marine snow'.

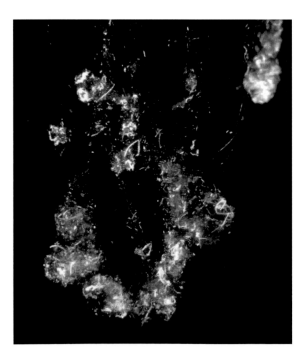

What is the fate of material released from the ice and plankton blooms

Of course not all of the material within the sea ice is living, and dead organisms and their remains accumulate within old sea ice. When ice melts the debris is released and sinks rapidly through the water. Likewise, not all the living organisms released from the ice go on to form blooms. A large percentage also fall downwards through the water. As described before (p. 88) there are large concentrations of polysaccharide material within the ice and this helps to stick algae and bacteria into clumps. Such aggregations are referred to as marine snow, and these fall at much faster rates through the water than the single phytoplankton cells.

A high percentage of this sinking material is broken down as it falls through the water. Bacteria produce enzymes that digest the cells and the contents. This means that as the material falls through the water it changes its biochemical nature – generally the nitrogen is used up faster than the carbon-rich compounds. As the bacteria and fungi break down the organic matter, they release nitrogen, phosphorus and other inorganic ions back into the water. As a particle or 'floc' of organic material falls it is colonised by more and more bacteria of different species, many of which stay tightly bound to the rich source of organic matter.

The faster a particle sinks, the less time there is for it to be broken down before it reaches the sediments below. In general, the greater the size of a particle the faster it sinks through the water, although sinking rates are also affected by factors such as biochemical composition: a phytoplankton cell with lots of lipid droplets (fats) within it is likely to sink at a slower rate than a cell with no lipid droplets. Cells with many appendages will fall more slowly than torpedo-shaped structures with no appendages to create drag.

Rapidly-sinking faecal pellets

The intense feeding activity at ice edges inevitably results in the formation of copious amounts of faecal material. Zooplankton and some protozoa produce faecal pellets that are bound by thin membranes. Therefore the remains of algae and often undigested algal cells are packaged into aggregations varying in diameter from a few to several hundred micrometres. The larger the grazer, the larger the faecal pellet, although not all faecal pellets are as durable as each other. Krill pellets, for example, are somewhat fragile and break up rapidly, whereas pellets produced by many of the small copepod species are far more robust and long-lasting.

Below
A faecal pellet full of sea-ice diatoms collected 400 m (1312 ft) under the ice.

The faecal pellets can sink through the water at fast rates varying from around 50–200 m (164–656 ft) per day. However, rates of up to 1500 m (4900 ft) per day have been measured. If a krill swarm is feeding under the ice or in an ice-edge bloom, a huge amount of faecal material is produced and sinks to deeper waters and the sediments below.

Naturally much of the material in the faecal pellets is broken up and partially digested. However, many of the common sea-ice diatoms have particularly strong cell walls (p. 87) and when there is abundant food a high proportion of the sea-ice algae in the pellets remain undigested. In this way large numbers of intact diatoms and sea-ice organisms are transported to the underlying sediments.

The sinking material made up of algal cells, bacteria and faecal pellets eventually collects on

the seafloor, and is known as phytodetritus. When there has been a massive fall of material, such as following an algal bloom, this material can form a loose fluffy layer several centimetres thick on top of the sediments. This is a source of food for organisms that feed by filtering out particles suspended in the water, or those that feed on organic matter between sediment particles, as well as being a rich source of organic matter that supports active bacterial assemblages.

Falls of phytodetritus can be so massive that they cause a disturbance to the small animals living in the surface layers of the sediments. This is because food falling is generally low so when there is a sudden huge downfall of organic-rich material the organisms are unable to cope. Foraminifers (different species to those in ice and water) are found in sediments in many parts of the world, where they rapidly take up and recycle organic matter. It has been shown that in the Arctic Ocean, where high falls of phytodetritus occur at ice edges, both released from the sea ice and after the ice-edge bloom, the foraminifer species diversity falls significantly compared to areas away from the ice edge.

Looking into the past

Researchers interested in studying the past cover of sea ice in polar regions can gain a lot of information by looking at the sea floor. If there is a high proportion of sea-ice organisms buried in an area of sediment, it is likely that these were released from overlying sea ice. By estimating the dates when the layers of sediments were formed it is therefore possible for palaeoceanographers to reconstruct the periods in geological time when sea ice covered those particular regions. It is not just the resistant diatom species that are used for these purposes, but other organisms such as siliceous nanoplankton species and the cysts produced by dinoflagellate species that are commonly found in sea ice.

The unique chemical characteristics of the sea ice tend to result in the sea-ice organisms having certain biochemical properties that are preserved within the organic matter when they are incorporated into the sediments. Therefore, it is not just the species themselves, but also the biochemical nature of the material within the sediments that provide the clues for reconstructing past sea-ice regimes. Carbon dioxide is often limited within the sea-ice matrix, and this results in the photosynthetic organisms that assimilate carbon dioxide during photosynthesis being limited by the gas supplies.

Measuring the fall of organic matter

Scientists study the flux of material falling through the water column by placing sediment traps at different depths. These devices are designed to collect the material raining down from above into sample bottles in which the organic matter is preserved. The bottles are retrieved and the contents analysed for chemical characteristics. They are also examined under a microscope to identify what is constituting the bulk of the material collected.

Sediment traps have been deployed in seasonally ice-covered waters for periods of time ranging from a few days to several years, and this has enabled researchers to build up a fairly detailed impression of the nature and magnitude of the flux of organic material to the underlying sediments. In both the Arctic and Antarctic there is little flux of material out of the ice during winter, and trap studies clearly show that the main annual fall of material starts after the ice begins to break up and melt. There can be a delay of up to three months from when the ice breaks up until significant material is collected in the underlying traps.

The peak fluxes of material tend to be in ice-free periods of the year, which for many parts of coastal Arctic regions and the Antarctic is just a few months of the year. Naturally much of the central Arctic is never ice-free and therefore the flux of organic material is particularly low all year round. In some parts of the Southern Ocean it has been estimated that up to 99% of the total annual

Left
Preparing small sediment traps to be deployed underneath the ice.

flux of organic material to the sediments takes place in just three months in the Antarctic summer between December and March. A large difference between the Antarctic and Arctic fluxes is that in the latter a significant proportion of the material (up to 30%) collected by sediment traps is not biological in nature but is of mineral origin. In contrast, in Antarctic sediment trap studies, the vast majority of the material collected is biologically derived.

Benthic communities

The organic matter reaching the sediment of the sea floor is a vital source of food for a vast array of organisms. Collectively these animals that live on and in the sediment of the seafloor are known as the benthos. A large proportion are suspension feeders, which means that they rely on food particles falling from above or material contained within the sediments being resuspended so that they can trap the particles and ingest them. These organisms are largely sessile (do not move) and include sponges, cnidarians, bryozoans, ascidians and echinoderms. They feed on particulate organic matter, extending from bacteria through to zooplankton several centimetres in length. Any well established benthic community comprises many different types of suspension feeder, making a three-dimensional community of organisms with widely ranging

Opposite top
A diverse variety of organisms make up benthic assemblages that feed on organic matter falling from surface waters above.

Opposite bottom
Starfish and nemertean (flat and unsegmented) worms scavenge on the seafloor in huge numbers.

Below
The bottle on the left is from a sediment trap deployed for five days under the ice. The brown colour is an accumulation of faecal pellets falling from the ice.

feeding strategies to attract and trap particles suspended in the water. By having a variety of organism types and sizes the whole spectrum of food particles falling on to the benthos is utilised.

Benthic organisms do not starve in the winter

In both the Antarctic and Arctic the benthic assemblages can be very rich. This is somewhat in contrast to patterns of production of organic matter in the surface waters, which is limited to just a few months of the year. It was long thought that these organisms simply fed on the short pulses of material from the surface waters, and then survived long periods without feeding. However, it now seems that this hypothesis is not true. Instead it is thought that during winter months the organisms feed on organic material resuspended into the bottom water layers when sediments are disturbed by water currents. In many regions this supply of food has been shown to be supplemented by organic material transported from shallow sites to deeper waters by near bottom water currents.

Interestingly there does not appear to be any uniform pattern in the coupling between benthic invertebrate reproduction and the seasonal pulses of food from above. In fact there are many cases when reproduction seems to take place at times of the year when the flux of food is minimal. An explanation for this is that the organisms utilise the high food influxes to build up their reproductive tissues, and then release their gametes later in the year when food is more scarce.

'Gigantism' and the oldest living organism on Earth?

One of the characteristics of many benthic organism groups in polar regions is the fact that they reach much greater sizes than their counterparts in warmer waters. Again this seems anomalous considering the apparent scarcity of food. Of course, small species are still present, and not all polar invertebrates are large. However, Antarctic sea spiders up to 40 cm (16 in) across are a hundred times the size of the common European sea spider. Isopods, such as *Glyptonotus antarcticus,* found throughout Antarctica, the Antarctic Peninsula and sub-Antarctic islands, can be up to 20 cm (8 in) in length and weigh 70 g ($2^1/_2$ oz). For comparison, isopods in other parts of the world may reach a maximum size of just several centimetres. Other 'giants' include sponges that are 2–4 m ($6^1/_2$–12 ft) tall and ribbon worms over 3 m (10 ft) long. It is thought that this 'gigantism' is brought about by a combination of factors.

Low water temperatures certainly slow metabolic rates to the extent that growth rates are slow enough to enable the organisms to live longer. Respiration rates are barely measurable in many of the benthic organisms by standard laboratory techniques.

Sponges are often the dominant organisms in benthic communities in the Antarctic. Some species, such as *Rosella nuda* (white volcano sponge), can be up to 2 m ($6^1/_2$ ft) high and weigh up to 500 kg (1102 lb). Researchers have estimated that the Antarctic lollypop sponge (*Stylocordyla borealis*) can live up to 150 years. The 30–40 cm (12–16 in) high *Rossellid* sponges are thought to be at least 300 years old and the largest, which are 2 m (6½ ft) tall, could be 10,000 years old. If true this would make *Rossellid* sponges the oldest living organisms on the planet. Naturally these ages and/or growth rates cannot be measured directly, and are mostly estimated from indirect measurements of the organism's metabolism combined with sophisticated mathematical models of growth rates. Therefore, there has to be a degree of uncertainty in these estimates, but it is clear that they are very old indeed.

Just how these organisms live so long is the subject of intensive studies. One possible theory is that their longevity is linked to the low cell metabolism at cold temperatures. During normal cell metabolism reactive oxygen products such as oxygen radicals and hydrogen peroxide are released. These potentially toxic products are thought to be influential in the ageing process. At the low temperatures in polar regions cellular respiration rates are low, and so the production of these harmful substances is also low, enabling organisms to live longer.

Below
Sponges are a dominant feature of many Antarctic benthic assemblages.

Above
Some Antarctic sponges are thought to be the oldest colonial organisms on Earth, growing at very slow rates.

Do oxygen levels control size?

It appears that the primary cause leading to variations in size between polar organisms and those from warmer climes is due to the difference in the dissolved oxygen content of the water. The physiological limit on the size of a particular organism is the amount of oxygen it can get into its blood, and in oxygen-rich waters this is largely dependent on the efficiency and length of the circulation systems supplying oxygen to the tissues. It is also true that in colder waters the oxygen demands of tissues are less, contributing to the possibility of a larger size.

Gigantism in polar waters is not restricted to organisms with highly-developed circulation systems, however. Pelagic organisms including ctenophores, copepods and pteropods all have polar species that are much larger than related species from temperate and tropical waters. These larger organisms have more tissue volume to which they must supply oxygen, but relatively less surface area with which to sequester the oxygen. Therefore, the higher oxygen contents of the low temperature polar waters are ideally suited to support these larger species.

Seaweeds under the ice

Many seaweed species (the macroalgae) are able to tolerate freezing temperatures, even those from temperate regions. Intertidal seaweeds, which are exposed during low tide, can tolerate periods of up to 12 hours or more of freezing at –20°C (–4°F) and below. Some seaweeds have been shown to survive freezing temperatures of –60°C (–76°F). As a rule of thumb, permanently submerged (subtidal) species are less tolerant of such cold temperatures. The same factors influencing microalgae survival in the ice apply to these seaweeds, and many polar species have been shown to have very high concentrations of compounds such as proline and DMSP that are thought to act as antifreeze agents. The production of reactive oxygen, such as hydroxyl radicals and hydrogen peroxide, is associated with freezing stress in seaweeds. Species that are better able to tolerate periods of freezing are more able to deal with damage by reactive oxygen. Just as in the sea ice, where it is thought reactive oxygen is a significant stress, these species have high levels of antioxidants and oxygen-scavenging enzymes.

Different tissue types are damaged to varying degrees by freezing, and growth regions are generally more susceptible to damage than older tissues. Seaweeds exposed to freezing temperatures in summer months are less tolerant of freezing than the same species in the winter. The speed at which freezing takes place is also crucial, gradual freezing causing less damage than shock freezing.

Left
Seaweeds in intertidal shorelines can become rapidly frozen when exposed at low tide.

Intertidal and subtidal seaweed flora

Seaweeds that grow in the intertidal zones of polar shores are greatly disturbed by the scouring effects of ice. For this reason many shores in polar regions are bare or populated only by relatively fast-growing ephemeral green seaweed species that spring up during the short period of the year when ice is not present. These stark landscapes contrast sharply with the luxuriant growth of intertidal seaweeds more commonly encountered on ice-free shores, and may give the impression that the abundance and diversity of seaweeds in polar regions is poor. However, in the subtidal zone, where the effects of ice scouring are less pronounced, about 100 seaweed species are found in the Antarctic and even more in the Arctic. Stands of large brown seaweeds, such as *Laminaria saccharina* and *Laminaria solidungula* in the Arctic and *Himanthothallus grandifolius*, *Ascoseira mirabilis* and *Desmarestia antarctica* in the Antarctic, form dense kelp forests similar to those known in temperate regions. These subtidal forests are important nursery grounds for fish and invertebrates, as well as helping to protect shorelines from erosion.

Some seaweed species, including the red *Iridaea cordata*, occur at very high latitudes of up to 77°S, where they grow in subtidal waters covered by sea ice for up to 10 months. Like other polar seaweeds that experience long periods of seasonal ice cover, they are generally highly adapted to low light conditions. With specialised photosynthetic pigments to exploit low light levels and cold-adapted metabolisms, the seaweeds are able to grow in virtual darkness.

Below
Rough waters laden with broken ice floes can effectively demolish underwater stands of seaweeds, and the debris washes up on the shore.

Left
Healthy, large brown
seaweeds collected from
under the sea ice of the
White Sea in Russia.

However, when the ice clears away, the clear polar waters, long day-lengths and high angle of the sun provide ideal light conditions for seaweeds which can reach growth rates comparable with species that live in much warmer waters. Just as for sea-ice algae described previously there are some complications. The low light adaptations mean that for some species there may be damage or inhibition of photosynthetic processes during the high light levels of spring and summer. These inhibitory effects are especially enhanced in connection with seasonally high levels of ultraviolet radiation that are associated with ephemeral holes in the ozone layer. Just as for the sea-ice algae, many of these seaweed species are able to produce protective pigments and MAAs when ultraviolet levels increase.

Below
Seagrasses and seaweeds encased in Arctic sea ice.

Seaweeds growing in the dark

Several species, including the brown *Laminaria solidungula* from the Arctic and the red *Palmaria decipiens* from the Antarctic, actually begin to grow during periods of darkness at the end of the winter, when the water surface is still covered by the ice. These species have not found a way of growing without light – instead they begin to grow by using the starches and other metabolites that they have built up in the previous year's growth period. The new tissues produced are ready to begin photosynthesis as soon as light becomes available when the ice breaks up. This kick-start maximises the growth period during the short summer months when light conditions are more favourable.

Ice and seagrass

Seagrasses are not found on Antarctic shores. However, they are important components of many Arctic coastal ecosystems where they grow to depths of 5 m (16 ft). In the White Sea, studies have shown that local populations of seagrasses in shallow waters can be decimated when weather patterns cause land-fast ice to be pushed offshore at low tide. The scraping of the ice causes severe damage to the surfaces of the sediment and the attached plants. Naturally this kind of disturbance will have ramifications for the viability of the seagrass meadows in the following spring and summer, and for the fish and invertebrates that use the seagrass beds as nursery grounds.

Differences between Arctic and Antarctic benthos

The benthic floras and faunas of the Arctic and Antarctic are quite different. In general the numbers of species in the major groupings of large benthic organisms are between 1.5 and six times greater in the Antarctic than in the Arctic. In fact the Antarctic species diversity is akin to the levels of diversity recorded for tropical regions.

One of the reasons for the greater species richness in the Antarctic compared to the Arctic is that the area of the Arctic Ocean is half of that covered by the Southern Ocean. Species richness is known to increase proportionally with increasing area, since larger areas have a greater variety of habitat types and can contain more individuals. Therefore just on size alone the Antarctic would be expected to have a greater diversity of species.

The differences are also compounded by the fact that there is a much higher degree of endemism (where a species is restricted to a particular region) among the Antarctic benthic flora and fauna compared to that of the Arctic. The degree of endemism varies greatly between different groupings of organism: up to 70% of the fish genera and 95% of fish species found in the Antarctic are endemic, whereas only 5% of polychaete genera and 57% of the polychaete species in Antarctic waters are endemic. Only about 5% of the seaweeds found in Arctic waters are endemic, compared to 30% of the Antarctic seaweeds. Interestingly, crabs, flatfish and balanomorph barnacles, which are common in the Arctic, are missing from the Antarctic fauna. Of course, these numbers are only approximate and change considerably as more detailed taxonomic research takes place and larger databases are established.

The differences in degrees of endemism of benthic organisms between the Arctic and Antarctic are largely due to the greater age and longer isolation of the Southern Ocean. The Arctic benthic floras and faunas have largely been established during the last two million years, and there are still close connections between the Arctic basin and the North Atlantic and northern Pacific oceans. These connections allow organisms to migrate to Arctic waters so that the Arctic benthos resembles neighbouring cold-temperate Atlantic or Pacific systems.

In contrast, the Southern Ocean was effectively cut off from the rest of the world's oceans 25 million years ago by the formation of the circum-Antarctic current. In the Southern Ocean the polar front and lack of shallow sea linkages effectively isolates the system. Over the millennia endemic species of algae and animals have evolved.

Damage due to ice scour

Coastal intertidal habitats are harsh places to live due to the extremes of temperature, salinity, desiccation and other stresses that the organisms endure when the tide goes out. One of the major stress factors is, of course, wave action, especially when the waters are full of sand or small pebbles so that the pounding waves are like industrial sandblasters. This is also true of waters that are covered by ice. In this case the blocks of ice, or more commonly small pieces of ice, scour the intertidal rock faces removing all but the very hardiest of organisms. In fact very little remains on an effectively ice-scoured shore, and typically these shores are characterised by fast-growing species of algae and animals that rapidly colonise the bare shores during ice-free periods, but which do not grow during winter months. Typically these are places where ephemeral green seaweeds such as *Enteromorpha* and *Cladophora* grow, which are colonised by a host of invertebrate species that live between the tufts of seaweed. Such species can rapidly cover a bare shore in less than two weeks under favourable growth conditions.

The ice-scouring effects are not restricted to exposed shores where waves pound ice pieces onto the shore. On coasts where there is significant tidal amplitude, large pieces of sea ice can scrape the shoreline every time the tide

Below
Ice-scoured shorelines on the coast of Finland.

goes in or out. On many shores an icefoot often forms – a narrow fringe of ice attached to the shore that is unmoved by tides. When this develops, although the shore is continually encased in ice, at least the disturbance potential of the ice scraping up and down shores is greatly reduced. Where an icefoot develops the biology still tends to be confined to crevices and cracks in the underlying rocks.

On sandy or muddy shores it seems that the native organisms burrow into the sediments to avoid being damaged by ice. Many invertebrates, including bivalves *Mytilus edulis* and *Mya arenaria*, the periwinkle *Littorina saxatilis,* and barnacles *Balanus balanoides,* can actually tolerate being frozen into ice, their hard shells protecting them against the mechanical damage. However, sometimes whole mussel banks can be lost if they become encased by ice floes and winds push the floes offshore, sweeping them into deeper waters.

Ice floes do not just scour the intertidal zones of shores. Where thick rafted ice is present, the movement of ice on tides and/or with waves can cause significant disruption to subtidal sediments and associated seaweeds and animals. Naturally the deeper one goes, the less disturbance there is from ice scour, although damage due to pack-ice scour is known to occur at depths greater than 50 m (164 ft).

Such ice scour can increase biological diversity. For example, when large perennial seaweed species (that effectively cut out light from below) are removed by ice scour, the resulting increased light reaching greater depths results in a greater diversity of annual seaweed species growing at the same site. However, an increased biodiversity with increased ice-scour effects is not a universal feature.

Keels of ridged sea-ice floes have been shown to influence the structure of benthic communities. A study in the Canadian Arctic showed that undisturbed assemblages were dominated by predators and suspension-feeding animals at depths of 3–15 m (10–50 ft). After disturbance by the keels of ice floes scraping across the sea floor, the assemblages changed and contained higher proportions of deposit-feeding and scavenging animal species. Likewise in the Antarctic it is thought that ice scour plays a large role in determining the composition of shallow water species in the Antarctic. It is also apparent that the localised devastation of benthic life by ice scour, and subsequent recolonisation, certainly helps in producing patchy distributions of sessile organisms on the sea floor.

The movements in ice and water due to tides is not all bad. The Inuit from the village of Kangiqsujuag are able to exploit mussel beds being exposed even during the winter when they are covered by ice. At low tide, the Inuit make holes in sea ice to reach to where receding waters have exposed beds of mussels. The fishermen effectively harvest the mussels with a roof of sea ice above their heads and have to endure meltwater from the overlying sea ice dripping on top of them.

Right
Ice floes in shallow waters are very effective at scouring the seafloor.

Below right
Icebergs can scrape along the seafloor damaging benthic assemblages that are gradually recolonised by benthic organisms returning to deep scour trenches.

Iceberg disturbance

As discussed at the beginning of the book, icebergs can travel huge distances following calving from an ice sheet and vary in size from a few hundred square metres to over 500 km^2 (193 sq miles). These move at considerable rates, and if large enough their keels will scour the sea floor. As the icebergs scrape across the sea floor, the devastation to organisms and the underlying sediments is immense. It is difficult to compare different sorts of ecosystem impacts, but iceberg scouring effects are thought to be one of the major large-scale disturbances of ecosystems on Earth. The damage caused is thought to be on the same scale of impact as cyclone damage to coral reefs, tree falls or fires in tropical and temperate forests and earthquake landslides.

The deepest plough marks due to iceberg scour have been recorded down to 600 m (1968 ft). Considering that at any one time there may be more than 300,000 icebergs in the Southern Ocean, the potential for tremendous seabed disturbance in coastal waters is great, especially considering that many of these are carried in the Antarctic coastal current for the beginning of their trajectories.

It has been estimated that for the entire continental shelf region of the Antarctic, up to 500 m (1640 ft) deep, there will be one disturbance per square metre every 340 years. In a similar approach in an Arctic study it was estimated that in some regions each square metre would be disturbed by iceberg scour once every 53 years. Recovery following such massive disturbance can be very slow indeed and some researchers have estimated that recovery can take from 65 to 500 years. Because of the very slow growth of many polar benthic species, areas disturbed in this manner are likely to be characterised by a continuous natural fluctuation between destruction and recovery.

The wide scales of disturbance intensity are thought to contribute to the overall high levels of Antarctic benthic biological diversity. The significance of iceberg disturbance is likely to increase with accelerated melting of ice shelves, although it must be stressed that the ice-shelf calving that receives much attention by the media often cannot be related to global climate warming. Iceberg scour effects are less prevalent in the Arctic due to there being far fewer icebergs compared to the Antarctic.

Mammals, birds and the ice

8

IN THE FROZEN WHITE DESERTS OF THE POLAR REGIONS IT IS STRIKING HOW A NUMBER OF MAMMALS AND BIRDS HAVE COME TO MAKE THESE MOST HOSTILE REGIONS THEIR HOME. THIS IS MADE EVEN MORE POIGNANT BECAUSE MANKIND – EVEN WITH THE MOST SOPHISTICATED OF TECHNOLOGIES – HAS NOT BEEN ABLE TO INHABIT POLAR REGIONS TO A LARGE EXTENT.

T HESE ANIMALS not only exploit the land, but also the pack-ice regions that cover the oceans. For many of these species the pack ice is a vital platform from which to hunt, give birth and wean young, or to escape from predators. There are many hostile environments on Earth, including the deserts and the deep oceans, but arguably the ice-covered regions are amongst the most severe.

Seals

It was the large numbers of seals in pack-ice regions that inspired many of man's ventures into the frozen seas, and sealing has been a major source of sustenance for people living on Arctic coastlines. Even within the Baltic and Caspian Seas there are seal populations that use the ice. The Baltic grey seal (*Halichoerus grypus*) and a Baltic subspecies of the ringed seal (*Phoca hispida botnica*) use Baltic sea ice for breeding on, although grey seals will use small skerries or islands when ice is absent. Caspian seals (*Phoca caspica,* sometimes called *Pusa caspica*) also use the ice that covers the northern Caspian Sea from January to late April in order to pup, nurse, mate and moult.

Over the last century there have been dramatic reductions in the seal populations of both of these seas. At the beginning of the 20th century there were thought to be 100,000 grey seals in the Baltic; by the 1970s there were just 2500. This decline was attributed to pollution with polychlorinated biphenyls (PCBs) in Baltic waters. PCBs are chemicals that were once used in electrical equipment and are particularly slow to breakdown when released into the

Opposite
Emperor penguin in pristine plumage.

A list of marine mammals and birds whose presence is characteristic of sea-ice covered waters

MAMMALS	Common name	Genus and species
Antarctic	Crabeater seal	*Lobodon carcinophagus*
	Leopard seal	*Hydrurga leptonyx*
	Weddell seal	*Leptonychotes weddellii*
	Ross seal	*Ommatophoca rossi*
	Antarctic fur seal	*Arctocephalus gazella*
	Minke whale	*Balaenoptera acutorostrata*
	Killer whale	*Orcinus orca*
	Sperm whale	*Physeter macrocephalus*
	Southern bottlenose whale or Antarctic bottlenose whale	*Hyperoodon planifrons*
	Arnoux's beaked whale	*Berardius arnuxii*
Arctic	Ringed seal	*Phoca hispida*
	Harp seal	*Phoca groenlandicus*
	Hooded seal	*Cystophora cristata*
	Bearded seal	*Erignathus barbatus*
	Walrus	*Odobenus rosmarus*
	Polar bear	*Ursus maritimus*
	Bowhead whale	*Balaena mysticetus*
	Minke whale	*Balaenoptera acutorostrata*
	Grey whale	*Eschrichtius robustus*
	Narwhal	*Monodon monoceros*
	Beluga or white whale	*Delphinapterus leucas*
	Killer whale	*Orcinus orca*

BIRDS		
Antarctic	Emperor penguin	*Aptenodytes forsteri*
	Adélie penguin	*Pygoscelis adeliae*
	Southern giant fulmar	*Macronectes giganteus*
	Antarctic fulmar	*Fulmarus glacialoides*
	Snow petrel	*Pagodroma nivea*
	Antarctic petrel	*Thalassoica antarctica*
	Blue petrel	*Halobaena caerulea*
	Wilson's storm-petrel	*Oceanites oceanicus*
	South polar skua	*Catharacta maccormicki*
Arctic	Northern fulmar	*Fulmarus glacialis*
	Ivory gull	*Pagophila eburnea*
	Ross's gull	*Rhodostethia rosea*
	Eider duck	*Somateria* spp.
	Oldsquaw duck	*Clangula hyemalis*
	Thick-billed murre or Brünnich's guillemot	*Uria lomvia*
	Black guillemot	*Cepphus grylle*
	Dovekie	*Alle alle*

environment. The population has now partially recovered to about 13,000. Likewise, pollution by pesticides and industrial pollutants in the Caspian Sea is thought to be responsible for the decline in the Caspian seal numbers from one million in the early 20th century to about 150,000 now. Severe hunting pressure has also led to the decline of the Caspian seals.

Ringed, harp and hooded seals

The most abundant seal species in the Arctic is the ringed seal that is found in fast ice and more solid pack-ice regions. These seals keep their breathing holes open in the ice using strong claws on their front flippers. Polar bears may wait next to a breathing hole to catch any emerging seals and for this reason each seal usually maintains several breathing holes so that it has alternative escape routes. The ringed seals are unique in that they give birth to their young in snow-covered (*subnivian*) lairs, and their young are born with white coats that help to camouflage them against a background of ice and snow. In order to avoid the polar bears, the young ringed seals also learn to swim at a very young age, entering the water when still being weaned, and have been known to dive to depths of 90 m (295 ft).

Harp seal pups are covered with white fur when they are born on pack ice that is not directly connected to land. Within 12 days of being born the pup is independent, ensuring that the young and mothers can avoid polar bears as quickly as possible. Even farther out in the pack ice hooded seals haul out to give birth to their young, which are fully weaned and independent from the mother just four days after being born.

Left
Within two weeks harp seal pups are independent of their mothers, ensuring they survive from predators such as polar bears.

The ice walkers

Walruses also use sea ice for hauling out, often using their distinctive tusks as ice axes, giving them one of their popular names: the ice walkers. They can also break through ice up to 20 cm (8 in) thick by ramming the ice with their heads. Their prey is mainly made up from molluscs such as clams, cockles and whelks growing in the benthos, and so they are restricted to hauling out on ice covering

Right
Walruses use ice floes to haul out onto, even using their tusks as effective ice axes.

Below
A Weddell seal hauled out onto the pack ice.

shallow coastal regions less than 70 m (230 ft) deep. However, walruses are known to eat ringed seals as well. Walruses do not keep open breathing holes, but instead rely on polynyas and open leads between ice floes for their access to the water. In summer they tend to haul out on ice, as well as land, since much of the summer pack ice drifts over water that is beyond efficient diving depths for the walruses. Bearded seals also haul out on drifting pack ice covering shallow regions since much of their prey is from the benthos.

Weddell seals

The equivalent of the ringed seal in the Antarctic is the much bigger Weddell seal, but in contrast to the northern latitude seals the Weddell seals do not have an obvious predator similar to polar bears. Therefore life, in at least that sense, is not so perilous for young Weddell seals. Weddell seals breed farther south than any mammal on Earth, and many haul out and breed on fast ice rather than in the pack ice where crabeater and Ross seals breed and raise their young.

Weddell seals spend much of the winter in the water, but have to maintain breathing holes in the ice, which they also use to haul out. In summer the seals

Below
A young Weddell seal cooling down in a pool of slush ice.

may spend less time in the water, hauling out for over 30% of the day. They keep the access points open using their teeth which they use to grind away the ice from the edges of the holes. Weddell seals tend to die relatively young, and this is largely because they wear their teeth away and eventually cannot feed properly.

They feed mainly on squid and fish, in particular the Antarctic silverfish (*Pleuragramma antarcticum*) of which they can catch a hundred or more within a single dive. Weddell seals can dive as deep as 700 m (2296 ft) and can stay underwater for over an hour. They have two main diving habits: shallow dives within the top 100 m (328 ft) of the water column and deeper dives of 200– 400 m (656–1312 ft). The shallow dives tend to be the longer and the seals can travel distances in excess of 10 km (6 miles) from the breathing hole on a single shallow dive. It also appears that the seals preferentially haul out during daylight and spend the night in the water, and also that the deepest dives tend to happen during the day.

The diurnal differences in diving behaviour may be related to the different distribution of the Antarctic silverfish at different times of day. The fish move into shallower waters under the ice at night, possibly to hunt the krill that are feeding on ice algae on the undersides of the ice – the stomach contents of silverfish that have been vomited up by Weddell seals have been shown to be full of krill.

Weddell seals give birth to their young in October and November, and the pups double their weight in the first three weeks on the mothers' 60% fat-rich milk. They are weaned after seven weeks and totally independent after two to three months, a striking difference compared to the much shorter weaning periods of the Arctic seals.

Below
A Weddell seal with a camera attached to its head in order to give scientists information about its feeding behaviour.

There have been many studies using data loggers attached to Weddell seals to track the depths to which the seals dive and distances travelled under the ice. Some devices have also been attached to record how frequently they open their mouths to feed. Most recently researchers have attached video cameras to the heads of Weddell seals to give a seal's-eye view of the under-ice world and how seals feed. When there is enough light it seems that some seals dive deep and then catch their prey by seeing them as silhouettes against the bright overhead ice, although other prey are taken from above.

Crabeater, leopard and Ross seals

Crabeater seals, which feed mainly on krill, are thought to be the most numerous seal species on Earth (estimates range up to 30 million), and they spend all of their lives among drifting Antarctic pack ice. The seals filter the krill out of the water using specially adapted interlocking teeth. It is thought that the distribution of this species is largely linked to the interactions between krill and sea ice (p. 118). They give birth to their young on ice floes, and can often be seen in large numbers crammed on to small floes where they wean their young and moult. Crabeater seals tend to live in more open marginal ice zones and can

Top left
Two crabeater seals 'equipped' with satellite transmitters that allow their movements to be tracked from space.

Bottom left
An injured crabeater seal hauled out on the ice to recover and get away from predatory orcas and leopard seals.

travel many thousands of kilometres on drifting ice floes. As ice melts in spring and summer, the populations of crabeaters move farther south with the receding ice edges.

A similar pattern is seen with the leopard seals. Leopard seals have a reputation as being fearsome predators, although up to 45% of their diet may actually be made up from krill. However, they do feed on penguins and in particular crabeater seal pups. Leopard seals frequently swim at the edge of ice floes or in waters close to penguin colonies waiting for their prey to enter the water, where the aggressive and agile seals are seen to be superb hunters.

Not too much is known about the seasonal distribution of the other seal species that frequents the Antarctic pack ice, the Ross seal. This rare seal feeds predominantly on squid, and it is not often seen because it hauls out on the thickest ice in fields of pack ice where ships and researchers seldom venture. However, it is clear that this species is highly dependent on sea ice for giving birth to its young and for moulting.

Fur and elephant seals

Antarctic fur seals are commonly seen on sub-Antarctic islands and in the Antarctic Peninsula region, but are very rare in even loose pack ice. They eat vast quantities of krill along with fish and squid, and so their major food source is one that is inextricably linked to sea-ice extent and dynamics. Along with the

Below
An elephant seal on the Antarctic Peninsula, yawning.

fur seals, the colossal southern elephant seal hauls out on the beaches of these ice-free islands. These animals can dive to depths of 1500 m (4900 ft) in a quest to feed on fish, particularly squid. They also generally avoid sea ice of any description, especially when young. However, satellite tracking studies of elephant seals from the Antarctic Peninsula have shown that adult female elephant seals can spend long periods within quite closed pack ice, where they are thought to feed on shoals of the Antarctic silverfish, which in turn feed on krill under the ice. The southern elephant seal populations are apparently increasing while the populations in other parts of the world are in decline, and it is possible that the food supplies, including an extensive feeding ground in winter under the ice, may be partly responsible for this.

Polar bears

Probably the most evocative of all the mammals associated with sea ice is the polar bear. Although polar bears do spend time on land, they are true marine mammals depending on the sea and pack ice for their existence. In fact polar bears, the largest of the non-aquatic carnivores, prefer to remain within the pack ice all year long if possible since the ice floes provide a platform from which the bears can catch their main prey of seals.

Polar bears are found throughout the ice-covered waters of the Arctic Ocean. The greatest majority of bears roam near pack ice that is thin or breaks open on a regular basis. Generally bears avoid heavily ridged, rough sea ice and thick multiyear ice, mostly because the densities of seals are low in these regions. The southern limit of polar bears is basically governed by the southernmost extent of sea ice. Several have been reported close to the North Pole, although generally it is thought that very few stray farther than 80°N since the ice is generally thick multiyear ice floes.

There are thought to be about 40,000 polar bears in the whole of the Arctic basin, although more cautious researchers would say that the correct figure lies somewhere between 20,000 and 40,000. These bears are not part of one large population, but rather divided up into several subpopulations. Bears have been tagged and recaptured since the 1970s, and more recently bears have been fitted with radio transmitters so that their positions can be logged using satellite tracking systems. The results of these studies show that from year to year individual bears remain in the same geographic region.

However, on shorter time scales bears have been shown to travel in excess of 30 km (19 miles) per day for several days in a row. Therefore within a year many polar bears may travel many hundreds, if not thousands of kilometres. These large roaming distances are, of course, extended because the bears are travelling on a moving platform of ice that is blown by wind and carried on ocean currents. Tracking studies have clearly demonstrated that the bears are

Above
Polar bears roam huge distances over Arctic pack ice in search of food.

not roaming aimlessly but that they know exactly where they want to go.

Although much of their journeys are completed on top of the ice, polar bears are very strong swimmers, and can swim distances in excess of 100 km (62 miles) at a time. In fact polar bears are sometimes observed several hundreds of kilometres offshore, probably because the ice floes they were travelling on had melted from underneath them. They use their large forepaws as effective paddles to swim with, while their rear paws act as stabilisers or rudders.

Roaming a long way increases pollutant accumulations

Bears from Svalbard and the Barents Sea roam huge territories – up to 270,000 km² (104,247 sq miles) – and recent studies have shown that these bears accumulate significantly higher concentrations of polychlorinated biphenyls (PCBs) in their fat, blood and milk compared with bears confined to smaller coastal territories. The bears that are covering large distances have to consume significantly greater food reserves than the bears with smaller ranges. The PCBs are contained within the prey, and because the polar bears need to consume more prey they consequentially consume more of the pollutants.

Polar bears and reproduction

Polar bears largely lead a solitary existence, and bears may have to travel considerable distances in April to find a mate. Females start to feed in excessive amounts after mating in order to gain weight as quickly as possible. Female bears typically weigh 150–175 kg (330–386 lb), but before they give birth to their cubs they may have laid down fat stores so that they weigh up to 450 kg (992 lb). In late August to October the females move to land where they make maternity dens. On entering the den they have no opportunity to feed again until the following spring. These dens are made in the hard snow of snow drifts or snowbanks, and some dens may be as far as 100 km (62 miles) inland from the coast, although typically they are within 50 km (31 miles). Maternity dens can also be dug in earth banks within the permafrost. The dens tend to be between 2–3 m³ (71–106 ft³) and may have an exit hole 2 m (6½ ft) away from the den itself. Cubs (normally twins) are born in November and are nursed by the mothers with fat-rich milk until late February or March. When born the cubs weigh less than 1 kg (2 lb), and when they emerge from the den they weigh 10–15 kg (22–33 lb). Cubs remain with their mothers up to $2^1/_2$ years and so females normally only have cubs once every two or three years. The mortality rate of cubs can be as high as 40% within the first year, but if the bears make it to adulthood they can be long lived reaching 30 years old.

Polar bears and feeding

Although polar bears are reported to eat a variety of food – including beluga whales, seaweed and grass – seals by far make up the greater part of their diet. Seals haul out of the ice using breathing holes and polar bears have learned to exploit this. The bears' hunting and stalking techniques are geared to catching the seals before they can escape through the hole to the relative security of the water. However, bears are also able to stalk seals from the water, swimming towards them and exploding on to the ice to grab the seal. The bears are also able to use surface melt features in spring and summer ice to swim mostly submerged so that they approach their prey unnoticed. Bears also wait for long periods of time beside seal breathing holes, knowing that the seal has to surface in order to breathe.

When snowdrifts cover breathing holes ringed seals excavate small caves or lairs above the holes. These lairs offer superb camouflage from hunting polar bears and ringed seals give birth to their young there. However, polar bears have such an acute sense of smell they can track down the young in these lairs through the snow. The bears actively hunt the seal pups in spring because young seal pups have a very high fat content. It is thought the bears may be able to detect a seal covered by up to a metre of snow and up to a kilometre away.

Whales

Whales are the largest creatures to roam the oceans and at certain times of the year the polar regions are rich feeding grounds. However, as evocative as these animals are, it is interesting to note that the total biomass of bacteria in the Southern Ocean is around 3×10^7 tonnes, whereas that of whales is only around 8×10^6 tonnes. Even a single krill superswarm extending over 450 km^2 (173 sq miles) was estimated to have a biomass of 2.1×10^6 tonnes. Therefore, although the whales are certainly the most conspicuous of organisms in the polar waters, it is always worthwhile putting their ecological role into context by comparing them to the smaller constituents of the food web.

The toothed whales

Killer whales are found in the Arctic and Antarctic, where they hunt in groups (pods), feeding mainly on other whales and seals. In the Antarctic, two types of

Right
Killer whales or orcas hunt
within the pack ice.

killer whale have been described, although it is unclear whether this distinction is strictly appropriate. The white form, which feeds on penguins and mammals (seals and other whales) is found in more open waters and loose pack ice. In contrast the yellow form feeds mainly on fish and is found deeper in the pack ice. The 'yellow' colouration is due to a covering of diatoms.

It is generally thought that killer whales move out from pack-ice regions during winter, or at most keep to marginal ice zones. However, there have been several reported sightings of killer whales within winter pack ice. In the Antarctic on one notable occasion in August, 60 killer whales were spotted together with 120 minke whales in pools of open water that were cut off from the open sea by 65 km (37 miles) of compacted sea ice. On another occasion in August a group of 40 killer whales of mixed ages were spotted in leads of water, 400 km (249 miles) south of the ice edge. The presence of a calf within this group indicates that the whales may have given birth in these sea-ice covered waters. There have also been sightings of killer whales in Arctic winter sea ice off west Greenland and western Alaska.

Of course, the prerequisite for all whales in ice-covered waters is that there are sufficient areas of open water for the animals to surface and breathe. This is why so few whales are found deep within the pack ice during winter, and why whales typically migrate to sea-ice covered regions from lower latitudes in summer months. However, where polynyas and areas of open water persist it is possible that whales can survive.

By feeding in pods the killer whales are able to tackle large prey such as other whales. They have been observed making coordinated leaps into the air, smashing up ice floes as they land and tossing seals on the floe into the water.

Sperm whales are another toothed whale species found in the outer margins of the Antarctic sea-ice zone. These larger whales feed almost exclusively on squid and can dive for periods of up to an hour to depths of up to 1000 m (3300 ft), which is where the larger squid species live. A much less common toothed whale in the Antarctic is the smaller southern bottlenose whale, which is found within regions of pack ice. These elusive whales can form groups of 10 to 20 individuals and have been seen to dive under sea ice for up to an hour, probably feeding on large shoals of fish such as the Antarctic silverfish or Antarctic cod.

Arctic toothed whales

Narwhals are probably amongst the most unusual looking of the toothed whales. All narwhals have two teeth in their upper jaw. After the first year of a male narwhal's life, its left tooth begins to grow into a long spiral tusk that can be up to 3 m (10 ft) long. The hollow tusks are usually twisted in a counterclockwise direction. The tusk's function is uncertain, although it is not used to catch the narwhal's prey of fish and squid. It is thought that narwhals can dive to depths greater than 1500 m (4900 ft), which makes them the deepest divers of all mammals. During winter they tend to move from coastal shelf regions (where they spend the summer) into deep Arctic basin waters where they are thought to be feeding on deep-sea squid. They are closely associated with the Arctic pack ice throughout the year and utilise leads and polynyas for moving through the pack ice in winter.

Belugas or white whales are also found deep within Arctic pack ice. They are able to move up to 80 km (50 miles) a day through thick pack ice, again utilising leads as places to breathe. Congregations of many belugas have been shown to keep areas of ice open by vigorously swimming at the surface of the water, continuously stirring up the water so that the ice cannot form. However, when they do this they become vulnerable to being caught by polar bears. Belugas can also dive to depths greater than 500 m (1640 ft) feeding on fish.

Left
The Arnoux's beaked whale, which is often mistaken for the southern bottlenose whale, is a rare sight in Antarctic pack-ice regions during summer.

Above
Beluga or white whales surfacing in an Arctic lead to breathe.

Right
Vast quantities of krill are needed to satisfy the whales, seals and penguins that rely on them for food.

Both belugas and narwhals do not have dorsal fins. This enables them to swim close to the underside of ice floes as well as enabling them to break thin ice with their backs.

The baleen whales

Bowhead whales can also break ice up to 18 cm (7 in) thick, aided by these whales having an elevated blowhole that is pushed through the ice to breathe. It is thought that the bowheads seek thinner ice or ice with breathing holes. It has been proposed by some scientists that minke, bowhead, belugas and narwhals all remain deep within pack ice in order to avoid, or at least reduce the potential of, predation by killer whales. Certainly migrating into regions of closed pack-ice is a risk for all of these species, and even though they can all break thin ice, there is a danger of becoming stranded deep within the pack ice with limited breathing holes.

Bowheads are baleen whales, which means they have baleen plates made from keratin that act like filters to sieve out the zooplankton on which these giant whales feed. The bowheads feed predominantly on copepods, and migrate northwards in summer through leads in the ice. They filter vast quantities of water since they need to catch about 3% of their body weight in food every day. For a 100 tonne animal, that means 3 tonnes of copepods a day.

Baleen whales also do not have dorsal fins, an adaptation that allows them to break ice. Likewise, minke whales have very small dorsal fins located near the tail, and this species can also break thin ice. Minke whales are found in both the Arctic and Antarctic, although they are different populations. They are the only baleen whales to move regularly deep within the pack ice of the Southern Ocean, some individuals staying within the loose pack ice edges the whole year round. These feed mainly on krill, whereas in the Arctic minke whales will be feeding mainly on copepods and amphipods.

Grey whales migrate over 10,000 km (6214 miles) from their breeding grounds off Baja California in summer to the open waters of the Arctic. Here they feed on the many tiny animals that live within the sediments on the benthos. Therefore, the length of the ice-free season is fundamental for grey whales, since they do not spend much time even in the least dense pack ice.

Other large baleen whale species travel huge distances in austral summer to feed on the dense krill swarms in the ice-free regions of the Southern Ocean. These include humpback, southern right, sei, fin and blue whales. These species are rarely seen close to the pack ice, and if so only at the very outer margins of receding ice edges. They feed predominantly on krill swarms, and as with other krill-eating birds and mammals, the interannual variations in krill recruitment and biomass greatly influence the feeding potential of these species.

Left
A minke whale surfaces between Antarctic ice floes.

Whale populations decimated by whaling activities

Of course, it was the migration of whales to the Southern Ocean feeding grounds that encouraged the whaling industry of the early to mid-20th century. However, these activities often overshadow the devastating whaling activities in Arctic waters, where by the early 18th century certain whale species had already become scarce. The numbers of whales in Arctic waters were further reduced after the mid-19th century with the advent of harpoon guns and efficient steam-driven whaling boats.

By the beginning of the 20th century the whaling industry in the Arctic was waning, and so the newly discovered whale stocks in the Southern Ocean were the obvious next focal point for the whalers in their search for whale blubber and oil. The first whaling stations were opened on South Georgia in 1904, and by 1916 over 7000 whales were caught per year. Between 1925 and 1936 the numbers of whales killed per year rose from 14,000 to 46,000. Interestingly, the minke whales were not as heavily hunted as the other species and it is thought that minke whales are now the most abundant whale in Antarctic waters. By the end of commercial whaling in the 1960s it is estimated that only 1% of the blue and humpback whales were left.

Below
Lichen-covered whale bones – poignant remains of the Antarctic whaling industry.

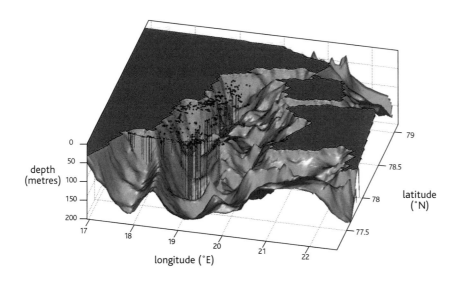

Left
Plot of water temperature and water masses logged on devices carried by beluga whales diving below sea ice in an Arctic fjord.

Below
A device to measure position, salinity, temperature and depth strapped to the back of a beluga whale which can travel under ice where scientists can never get to.

Using whales as oceanographers below the ice

On a more positive note, an intriguing development in recent years has been the use of whales to carry oceanographic equipment to measure the physical characteristics of water underlying ice. By using whales that can spend considerable amounts of time under ice or at ice edges, oceanographers can gain insight into regions that hitherto have been inaccessible. In a sense the whales are biological automated underwater vehicles (AUVs) (p. 84), although the whales cannot be preprogrammed to move in any particular direction.

Two wild, free-ranging belugas were equipped with devices to measure depth, salinity and temperature in an Arctic ice-covered fjord in Svalbard. The data was logged as the whales dived to depths of 180 m (590 ft), and when the whales surfaced to breathe the data was automatically transmitted to satellites and then to operators on land, where the data was collated and mapped. Although the sensor on one of the whales failed, data from the other was returned for 63 days, enabling the distribution of water masses under the ice to be described over an area of 8000 km² (3088 sq miles).

There is a huge potential in using satellite-transmitting data-loggers attached to marine mammals, especially in ice-covered regions where operating ships is both expensive and logistically difficult. Presumably as technology advances and devices become more reliable these techniques will be utilised more extensively. On the downside, there is little that can be done to prevent the data-loggers from falling off the oceanographic mammal, and clearly studies will be compromised if the organism dies or migrates to waters outside of the investigation area.

Penguins and other birds of the pack ice

Besides the polar bear, the flightless penguins best characterise the inhabitants of frozen pack-ice regions. Unlike the polar bear, they live in the Antarctic. There used to be a somewhat similar flightless bird in the Arctic that foraged on the edges of the pack ice, the great auk (*Alca impennis*), although this had been hunted to extinction by 1844.

Four species of penguin breed on the Antarctic continent, although it is only the emperor and Adélie penguins that can truly be said to be associated with sea ice. Other species including the king, chinstrap and gentoo penguins, are normally found only in the more northerly parts of the Antarctic Peninsula. In fact the Adélie and emperor penguins are referred to as ice obligate species, which implies that if there was no sea ice they would probably die out.

The emperors

Estimates of the numbers of breeding pairs of emperor penguins vary from 100,000 to 200,000 in about 30 to 40 breeding colonies dotted around the continent. They use land-fast ice for incubating their eggs and raising their young. In February and May female and male emperors travel back to their breeding colonies after a period of extensive feeding at the ice edges. By this time they have laid down extensive fat reserves through gorging themselves on krill, amphipods, squid and small fish.

The female lays an egg in May and passes it to her mate, before going back to the ice edge to feed. The non-feeding male lives off his fat reserves through the long winter (100 to 120 days) while incubating the egg in a specialised brood pouch just above his feet. The birds keep the egg within the warm pouch and off the ice surface using their feet, and the temperature difference between the inside and outside of the pouch can be up to 80°C (144°F), when the coldest air temperatures of –60°C (–76°F) are reached. During the cold winter a male emperor penguin can burn off about 200 g (7 oz) of weight a day, and the colony of males huddle together in tightly packed groups to conserve heat and energy, taking turns to occupy the coldest places on the edge of the group.

In July or August the eggs hatch and females return from the ice edge, often a trek of over 100 km (62 miles) or more. Remarkably mates locate each other, no mean feat in a colony of over 10,000 birds, and the female takes over looking after the chick, feeding it on regurgitated food. The males then walk to the ice edge to feed and bring back food for the growing chicks. After two months the chicks are so big that they cannot be kept warm by the parents, but instead huddle together with other chicks to keep warm.

By December, the adults stop feeding the chicks and the chicks have to move to the ice edge to enter the water for the first time. By this time they have

Below
Antarctic Adélie (left) and emperor penguins.

lost their juvenile feathers and have grown their adult water-repellent plumage. The parents will also have spent some weeks moulting and replacing their feathers. While they are doing this they cannot enter the water and so sea ice provides a very necessary platform on which to sit out this vulnerable period.

Adélie penguins

The Adélie penguins are the most numerous of the Antarctic penguin species, and estimates are around two million breeding pairs. Although these small penguins do feed on small fish, the main part of their diet is thought to be krill. Adélie penguins do not nest or raise their young on ice, but instead nest on land in spring and summer.

In contrast to the emperor penguins, they spend the winter at the most northerly edges of the pack ice up to 700 km (435 miles) away from the land. Before the pack ice breaks up in late winter and early spring the adults walk and swim back to their breeding colonies on the land. Therefore, the winter migration of adult birds to the ice edge and back may involve journeys of several thousand kilometres through the pack ice, and some birds have been tracked by satellite as travelling 5000 km (3100 miles). Their breeding colonies

Above
'Punk' style young Adélie penguin shedding juvenile feathers.

are always situated in locations that will become ice-free once the chicks have hatched, ensuring that the young can feed before travelling north for the austral winter months.

While the young wait in the colonies, the adults must feed them, and typically travel up to 30 km (19 miles) to find krill. However, distances up to 235 km (146 miles) have been recorded for some individuals. The Adélie penguins can dive for several minutes to depths of 175 m (574 ft). However, typically they dive to depths of less than 50 m (164 ft), which is where krill are normally found. Emperor penguins can dive to even greater depths of around 265 m (869 ft) when they are hunting squid.

Even though both species can dive to great depths, both Adélie and emperor penguins feed on the organisms living in the waters immediately under ice floes, especially krill, amphipods and small fish. These two species can hold their breath for longer periods than similar sized penguins, and they can swim for appreciable distances underneath the ice. Like the emperor penguins, Adélie penguins also moult their feathers by sitting on icebergs and ice floes after feeding and building up the fat reserves that enable them to be out of the water, and therefore not able to feed, for several weeks.

Petrels and other birds of the pack ice

A number of petrel species are clearly associated with sea ice. The snow petrel is even mentioned in maritime guides for sailors as a good indicator for sea-ice conditions and, like the two penguins discussed above, this is an obligate ice species. The entirely white plumage of the snow petrel acts as camouflage as the bird sits on the edges of ice floes waiting for fish and crustaceans to venture up from under the floe edge. When this happens the petrel will dive to catch its prey. Snow petrels will take advantage of any available food source, however, and they have been observed picking at the faecal matter of Weddell seals around breathing holes.

Snow petrels do not nest near pack ice – they prefer rocky ledges or cliff faces. Snow petrels have been found nesting more than 180 km (112 miles) from the coast on mountain summits (nunataks) that rise out of the icecap that is covering the Antarctic continent. This is even more remarkable considering that the adults must fly backwards and forwards to the sea ice to collect food for their chicks.

Snow petrels are perfectly camouflaged as they sit on the edges of ice floes waiting for crustaceans and small fish to catch.

Top
Ivory gulls feed from the edges of Arctic ice floes.

Above
Thick-billed murres dive under ice floes in search of food.

Ross's and ivory gulls in the Arctic use similar strategies to the snow petrel, where their white colouration enables them to feed from the edges of ice floes on Arctic cod, other fish and amphipods grazing on the underside of ice floes. Just like the snow petrels the ivory gulls also nest in snow-free mountain outcrops. It is thought that snow petrels and ivory gulls seek these nesting sites so that they have a huge expanse of ice around them to protect them from predation by skuas, and foxes and polar bears respectively.

Another common petrel in the Antarctic pack ice, especially in marginal ice zones is the black-and-white Antarctic petrel. In contrast to the snow petrel this species flies quickly along water edges and dives into the water to pursue prey underwater. The bird can swim underwater using its wings for propulsion. Likewise the Arctic black guillemots and thick-billed murres feed by diving under ice floes to search for crustaceans and fish.

There are many other bird species, such as the fulmars, duck species and other petrels that are found in ice-covered waters. These do not have behaviour patterns that are linked to sea ice; it seems that these birds tolerate ice-covered areas rather than exploiting the ice. In the Arctic especially, many of these species are found to congregate around polynyas and other areas of open water.

Mammals, birds and the changing sea-ice extent

Naturally any change in sea-ice conditions as a result of global climate change will have a considerable impact on the mammals and birds that live on and around the ice. From the predictions of current sea-ice trends it appears that the most profound decreases in ice extent will be encountered in the Arctic, in particular the extent of summer sea ice in the northern hemisphere.

Predicted decreases in summer sea ice and decreasing concentrations of multiyear sea ice will result in polar bears migrating farther north to forage for food. However, increasingly the pack ice will be cut off from land for longer periods of time. In the western Hudson Bay earlier break up of sea ice is forcing polar bears on to land earlier in the year thereby preventing them feeding fully on seals and building up sufficient fat stores with which to survive the winter.

Polar bears usually return to the same area for overwintering and often will utilise the same maternity den several times. However, warming of some permafrost regions have resulted in such dens collapsing, making the suitability of dens more precarious than has been the case up to now.

For whales that seasonally move into ice-free waters, presumably a decline of sea-ice extent in the Arctic, especially thick multiyear ice, will encourage whales to move into regions of the Arctic basin that they have not frequented before. However, seal species that need to be able to haul out on the ice will be concentrated into ever decreasing regions of ice as it decreases in extent. This will not only affect the behaviour of the seals, but also the animals that predate them.

In the Antarctic, the overall ice extent does not appear to be changing as dramatically as in the Arctic. However, clearly changing ice distributions will have an impact on those species that depend on the ice for food and migration patterns. Variations in Adélie and emperor penguin and snow petrel numbers around the Antarctic have been monitored since the late 1950s. No conclusive trends can be made by comparing the numbers of these species with ice extent. This is partly because these species have a circum-Antarctic distribution, and most studies have been restricted to monitoring a restricted number of breeding colonies.

However, in general it is thought that more ice means a greater survival or breeding success for these three bird species. However, increased ice extent has been demonstrated to cause increased mortality of male emperor penguins because of the increased distances the birds have to travel to feed after incubating the eggs in winter. Increased ice may also result in increased mortality of adult snow petrels since there will be fewer areas of open water, thereby reducing access to the water they need for feeding.

Clearly the effects of sea-ice extent and krill (p. 118) will be central to numbers of these species, since the krill is a major food source. Also the extent of sea ice will greatly influence the length of winter migrations to the ice edges for Adélie penguins, and there are arguments that in severe ice years there may be a long delay before young penguins can enter the water, reducing the numbers of individuals that can feed enough before the seawater freezes over.

Studying the pack ice

9

DESPITE MODERN SHIPS, SATELLITES AND AUTOMATED MEASURING DEVICES, WE KNOW LITTLE ABOUT THE PACK ICE IN WINTER WHEN ACCESS IS SEVERELY LIMITED. THERE IS MUCH TO BE DISCOVERED ABOUT THE COMPLEX PHYSICAL, CHEMICAL, BIOLOGICAL AND GEOLOGICAL CONSEQUENCES OF THESE FROZEN REGIONS. WE MAY EVEN BE ABLE TO USE THIS KNOWLEDGE WHEN WE LOOK FOR LIFE IN EXTRATERRESTRIAL SYSTEMS.

ONE OF THE MOST STRIKING DIFFERENCES between the Arctic and Antarctic is the degree of human activity. Mankind has had very little contact with the Antarctic before the last 200 years, and it is only really in the past 50–60 years that there has been a noticeable human presence in the region. Even now, the population of the whole Antarctic continent in Antarctic winter is limited to approximately 1000 scientists and support crews in less than 25 stations. In the summer these numbers swell to about 15,000 people of whom approximately two thirds are tourists. Both scientific and tourist activities are largely restricted to very small stretches of the coast, and by far the greater proportion of the Antarctic coastline is not visited.

In contrast, the Arctic has a population of around four million indigenous people (about 30% of the total Arctic population) that have lived in regions of Alaska, Canada, Greenland, Siberia and northern Scandinavia for many thousands of years. In fact, the recent carbon-dating of artifacts from a site on the Yana river in Siberia indicate that humans possibly occupied the region 30,000 years ago. These days the region is generally sparsely populated, and settlements vary from a few industrialised cities through to small nomadic communities comprising no more than a few families.

Traditional occupations for many of these groups are based on fishing, hunting, trapping for fur, reindeer herding and crafts. Clearly an intimate understanding of sea ice and seasonal changes dictated by frozen coastal waters is central to many of these activities.

Opposite
Scientists being deployed onto an ice floe using a 'mummy chair'.

Right
There are numerous bases or stations dotted around the Antarctic continent from which scientific research is conducted.

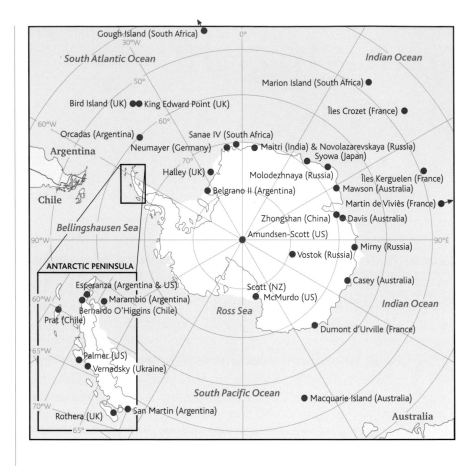

Little ice age and human habitation

There is good evidence that the effects of climate change on the pack ice has greatly influenced populations of Arctic dwellers in the past. Greenland was first inhabited by the Norse in around AD930 where they established settlements and a subsistence lifestyle along the south and south-west coasts. A rapid cooling at the beginning of the little ice age in the early 1300s caused drifting pack ice to expand over the North Atlantic, which hampered and eventually halted navigation between Greenland and Iceland. As it got colder and the pasture and farming lands shrank under the advancing ice and snow, the inhabitants suffered a painful annihilation, and contact with settlements ceased around 1450. Interestingly the Inuit of the region survived the worsening conditions, possibly because of their superior hunting skills and techniques. It is also possible that Norse settlements were overrun by hostile Inuits. This example highlights the fact that the impacts of climate change are likely to be felt first in extreme regions such as the Arctic.

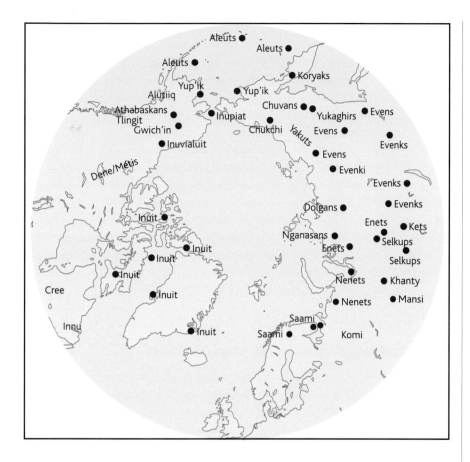

Indigenous peoples

The word Inuit means 'the people' and is used to describe a diversity of cultures spanning over more than 3000 km (1864 miles) of coastline, from the Kalaalit of Greenland, Iñupiaq of Canada and Alaska, Alutiiq in Alaska and Yup'ik of Alaska and Siberia. There are about 40,000 Inuit in Alaska, 30,000 in Canada, 40,000 in Greenland and about 1000 in Siberia.

The Inuit are not the only people to exploit the Arctic pack ice: the Aleuts inhabit the Aleutian Islands and parts of western Alaska. They are a seafaring people and commercial fishing and hunting of marine mammals, including whales, are key economic activities. The Chukchi people live in remote villages along the Bering Strait, Siberia and the Arctic and Pacific coasts of north-east Asia. The Chukchi people are divided into two main groups, the first of which depends on reindeer herding. The second major group lives in coastal regions traditionally hunting marine mammals and fishing. The Saami people live in Russia, Sweden, Finland and Norway, and reindeer herding is the main economic activity linked to these populations. However, fishing and hunting are major activities for coastal-dwelling Saami. In Russia, the Nenet people are traditionally nomadic, moving large distances along the Kanin Peninsula in Siberia. They also depend on reindeer herding, although in summer they hunt in coastal regions. The Athapaskan people stretch from Alaska, the Yukon Territory, Northwest Territories of Canada and into British Columbia. Fishing and hunting of marine mammals are central to many Athapaskan communities.

The lifestyles and traditional cultures of all of these groups have been greatly changed by outside influences in the past 100 years. Increasingly impacts of mining, timber, hydroelectric, gas and oil industries exert tremendous pressure on traditional lifestyles as well as on the environment in which many of the traditional activities are based. However, despite all of these developments, it is in these northern peoples that the understanding of and respect for the seasonal dynamics of pack ice is most finely honed. This knowledge is intertwined with a sophisticated understanding of the behaviour and migration patterns of mammals, birds and fish that after all were the main reason for venturing on to the pack ice.

The history of pack-ice exploration

It is thought that people have inhabited Arctic coastlines for tens of thousands of years, and it is clear that despite being a hostile and largely uninhabitable region of the world, frozen seas have attracted explorers since the earliest days of recorded history. 'Mare concretum' and 'Mare immotum' are terms thought to describe frozen, solid seas used by the Roman Pliny the Elder (c.AD23 to 79) and Strabo from Asia Minor (c.63BC to c.AD21). Even earlier, Pytheas of Massalia (now Marseilles) is thought to have encountered sea ice in 350 to 320BC in a place called Thule. It is not clear where Thule actually was, although Iceland, Denmark, the Shetland Islands, Norway or the Skagerak have all been suggested as possible locations. Ancient Greek and Roman geographers believed Thule to be the most northerly region of Earth, and today Thule is the name of a settlement in NW Greenland, the most northerly human community.

'Mare concretum' was a term also used by Irish monks to describe pack ice on journeys made to Iceland in AD795. The medieval text *Navigatio Sancti Brendani Abbatis* tells of a journey of St Brendan (AD489 to 570) in a leather-covered boat from Ireland to North America, where sea ice is clearly described as well as icebergs that are reported as being 'floating pillars of crystal'. The saint's journey was repeated in 1970, using a similar leather-covered boat, which proved that it was possible and that the descriptions in *Navigatio* were accurate. Interestingly the 20th-century explorers encountered perilous fields of sea ice and icebergs in the same regions that St Brendan is supposed to have encountered them.

In 1070 the German Adam of Bremen produced a monumental text on the peoples of the north called *Gesta Hammaburgensis Ecclesiae Pontificum*. He described sea ice and included descriptions of Greenland, Iceland and Vinland – the name given to part of North America when discovered by the Viking Leifur Eiríksson in the early 11th century. Putting the story of St Brendan to one side, the Viking Bjarni Herjólfsson is widely believed to be the first European to view mainland North America while sailing from Iceland to Greenland in AD986. Clearly all these seafarers will have known much about the pack ice in these waters.

The 15th and 16th centuries

Olaus Magnus Gothus (1490 to 1557) wrote *The History of the Nordic Peoples* (1555) as well as producing the first accurate map of Scandinavia, *Carta Marina*, in 1539. He placed the 'Mare Glacial' (frozen sea), illustrated by a pile of ice floes, just off the coast of Iceland. He was very aware of the importance of sea ice to the people living around the Baltic Sea, and illustrated his map with pictures of people crossing between Sweden and Finland on sledges over the sea ice, and using ice floes to hunt and fish. His patron was Pope Paul III, who absolutely refused to believe that the ice around the Baltic could possibly support the weight of a man, let alone the horses and sledges depicted in *Carta Marina*.

At about the same time the first ideas about a southern equivalent of the frozen northern pack ice were published. *Terra Australis Incognita* appears in maps by Orontius Finaeus (1531) and Ortelius (1573). The ancient Greeks had imagined 'Antarktikos', the first mention being made by Aristotle in 322BC. Antarktikos was derived from 'opposite the bear', 'Arktos' being the great bear constellation of stars above the North Pole. It is possible that the first human to encounter the Southern Ocean was actually the Raratongan, Ui-te-Rangiara in the seventh century who, according to Polynesian legend, is said to have "sailed south to a place of bitter cold where white rock-like forms grow out of the frozen sea".

Below
Portion of the *Carta Marina* (1539) where the sea ice in the Baltic Sea is shown for transport of people and a platform from which to hunt seals.

The Northeast and Northwest Passages

Sea ice and icebergs were prominent features of expeditions by pioneering 16th-century explorers to the Arctic such as Martin Frobisher, John Davis, Henry Hudson, William Baffin and Willem Barents. However, it is clear that indigenous people and Russian traders made great strides in exploring coastal and river routes during the 15th and 16th centuries. Clearly the discovery of the Northeast Passage from northern Scandinavia across the north coast of Siberia, and the Northwest Passage through the north of Canada were great incentives for these ventures in a quest to speed up trade routes between Europe and the Orient.

The Dane, Vitus Bering is credited with discovering the last parts of the Northeast Passage as a result of the 'Great Northern Expedition' that started out in 1733 and finished in 1743. During this expedition sponsored by Russia, various smaller campaigns, coordinated by Bering, started from land and sea locations to complete the task that had frustrated many before them, especially many Russian traders. The whole venture involved nearly 10,000 men, many of whom died. One of the pioneers of Arctic exploration was the Finnish born Adolf Erik Nordenskjöld who was the first to navigate the Northeast Passage on the Swedish steamship *Vega*. He set off from Norway in June 1878 and arrived in Yokohama in September 1879, having spent 10 months in the winter ice.

The discovery of the Northwest Passage was to prove more illusive and even more costly in lives of explorers who set out to discover it. John Ross, William Edward Parry, James Ross and John Franklin, among others, all led unsuccessful 19th-century expeditions in search of the passage. Sir Robert McClure finally discovered the passage in the 1850s, although his crew had to traverse large parts of this on foot, because the sea ice was impenetrable by ship. The 900-mile east-west passage runs through the Arctic Islands of Canada from Baffin Island to the Beaufort Sea, and then into the Pacific through the Bering Strait, which separates Siberia from Alaska. Even after the Northwest Passage was discovered, it took another half century before a single ship sailed through it. The Norwegian explorer Roald Amundsen, the first man to reach the South Pole, made the passage between 1903 and 1906.

Early days in the Southern Ocean

The first recorded expedition to cross into the Southern Ocean was by Edmond Halley onboard HMS *Paramore* in 1699 to 1700, although they ventured only far enough to see icebergs, but not pack ice. In the 18th century numerous explorations pushed farther and farther south driven by the goal of discovering the 'Terra Australis Incognita'. A Frenchman Philippe Buache even supposed the Antarctic to have the same general structure as the Arctic: a land-locked polar sea with large rivers similar to those known in Siberia. However, even in 1759 Charles de Brosses described the Antarctic as being a colder place than the Arctic.

Between 1772 and 1775 Captain James Cook ventured into the unknown with the ships *Resolution* and *Adventure* to make the first realistic map of the Antarctic continent. They reached their most southerly point in February 1774 where their progress was halted by the pack ice. Cook and the naturalist on board, Johann Reinhold Förster, made many pertinent comments on sea ice and how wind and seasons must influence the distribution and growth of the ice. They even conducted experiments with freezing seawater, and realised that older, desalinated sea ice can provide a good source of drinking water. Cook even considered the fate of brine that forms when seawater freezes.

In 1819 to 1821, Captain Thaddeus Fabian von Bellingshausen made a circumnavigation of the Antarctic continent in the *Vostok* and the *Mirnyi*. This expedition is commonly cited as being the first to see the ice shelves of the Antarctic continent. Bellingshausen also made observations about the nature of the floating pack ice, although he wrongly supposed icebergs to be formed by freezing seawater below and snow from on top. He also conducted experiments to compare the freezing properties of seawater compared to freshwater. In 1846 the American James Eights made highly insightful observations on the transport of rocks and animals by drifting ice, and speculated about the effects of icebergs grounding on the sea floor.

Below
In 1773 Captain James Cook ventured into the unknown of the Antarctic pack ice.

The experience of the sealers and whalers

At about this time sealers ventured farther south in search of seals for fur and elephant seals for oil. They used ships tested in the Arctic, and it is thought that the first sealing ships ventured into Antarctic waters in the late 1780s. Soon ships from many nations were using Antarctic hunting grounds to supplement the trade from ransacked seal stocks in other parts of the world. James Weddell was a sealer who led a voyage (1822 to 1824) and attained the feat of reaching the southernmost point to that date in the Weddell Sea, which was named after him, as was the Weddell seal. He made many fundamental measurements of the pack ice and took air and water temperature measurements.

Weddell even made pertinent comments about the biology of elephant and fur seals, emphasising the fact that the slaughter of the animals could not go unchecked if there were to be sustainable yields of furs. This advice was not heeded and by 1830 fur seals had already reached such low numbers that the sealers turned their attention to elephant seals for oil. By 1870, elephant seal stocks had been so decimated that the industry was hardly viable.

Below
The sealers and whalers discovered much about the seasonal dynamics and properties of sea ice.

During this period, whaling was a global industry and as whale stocks dwindled in other regions the whalers had to turn to the Arctic and eventually the Antarctic for their prey. One of the most famous whalers, who like Weddell also conducted scientific studies, was William Scoresby. He sailed extensively in the Arctic producing many scientific papers and *An Account of the Arctic Region,* published in 1820. In this he talks about many scientific issues from ice-crystal formation, to plankton distribution and marine mammals. Many of his descriptions about seawater properties and movements were quite remarkable, and he correctly identified the phytoplankton and realised that these microscopic organisms fed the zooplankton that were ultimately the food of the great whales.

Into the 20th century

In the 20th century the studies of pack ice really took off, and in both the Arctic and Antarctic pioneering scientists made giant strides in the understanding of the polar oceans and the seasonal dynamics of pack ice. Ship design, and safe passages for shipping were still a major driving force in the Arctic, pushing this research forward. In the Southern Ocean a lot of the emphasis became caught up with the massive whaling interests that dominated the beginning to the mid-part of the century.

However, any more detailed chronicle of sea-ice investigations from the beginning of the 20th century to present day is far beyond the scope of this book. There are several histories of polar research that discuss the many expeditions and studies in great detail. Many of these journeys were not envisioned specifically to make sea-ice observations, although such was the enthusiasm and scientific prowess of many of the early pioneers that ships' logs, expedition reports and accounts are full of insightful information about sea ice.

The famous expeditions of Scott and Shackleton resulted in sea-ice studies being published. Wright and Priestley who were with Scott on his last Terra Nova Expedition in 1910 to 1913 published a pertinent work that discusses several sea-ice issues, including the first observations about platelet ice. In 1921, Wordie who was a member of the 1914 to 1917 *Endurance* expedition, led by Shackleton, published insights into sea ice and the natural history of the Weddell Sea pack ice. Not only did these men survive a remarkable journey across the Weddell Sea ice after their ship sank, they still made pertinent scientific observations along the way.

Above
The *Endurance* before she sank beneath the ice of the Weddell Sea in 1915.

Ships, powerful engines and satellite navigation

The study of pack ice has grown at a tremendous rate as advances in shipbuilding techniques enabled ships to be built that could traverse thick pack-ice fields, or survive being stuck in the ice for long periods of time. Most studies are still conducted from ships, and these are operated by national agencies that coordinate their efforts so that multinational groups of scientists can study the aspects of sea ice. The overseeing of these efforts is made by a number of international organisations that encourage new ventures and collaboration in polar research.

Today, research cruises to the Arctic and Antarctic pack ice still have a sense of adventure and the romance of working 'at the ends of the Earth'. However, they certainly do not have the same sense of danger and unpredictability that must have pervaded all aspects of life on board the early expeditions. Members of the crew and scientists alike must still respect the vagaries of ocean and ice, and can never become complacent about the harsh conditions within which they are working. However, they are able to do this having the latest of satellite-based navigation equipment, sophisticated real-time weather reports and satellite pictures of ice concentrations in the region. The most modern ships are engineered to work in the hostile conditions, and powerful engines can enable passage through ice that would have stopped all ships only 40 years ago.

Modern communication has also transformed the lives of those working with sea ice. Until fairly recently any expedition to the pack ice would mean that there would be only the most limited communication with home, if at all. These days there is easy e-mail telephone communication. Photographs and daily update reports are now a common feature of research cruises in both the Arctic and Antarctic pack ice. Via the internet it is possible for those sitting at home to see the current position of a ship, as well as being able to read a whole host of different types of data online, everything from water temperature and salinity through to chlorophyll concentrations in the water.

It is not just the transformation in ship design over the past 50 years that has revolutionised pack-ice research. There has been an even more remarkable revolution in the technology and tools available to scientists for making their observations and conducting experiments. Most ships contain empty laboratories before each journey, and the various scientists on board bring their own analytical equipment and sampling gear with them, which is often packed many months in advance and stored in containers on board the ship. Very quickly an empty ship can be transformed into a suite of laboratories where the most sophisticated of analysis can take place, anything from molecular biology, geophysical measurements of the sea floor below the ship through to atmospheric chemistry.

Left
Specially strengthened ice-breaking ships are central to pack-ice research.

It is seldom that all the scientists on board a research cruise will all be studying sea ice. Ship time is expensive and there is strong competition for the limited places on the ships. Therefore scientists from many disciplines will vie for time on a cruise to do their work. There maybe oceanographers, chemists, fish biologists and sea-ice workers, all wishing to deploy different instrumentation and use ship time. Naturally a lot of planning goes on before the ship leaves so that the different demands can be met in the most efficient way. For this reason many expeditions are planned several years in advance.

Access to the ice

Sea-ice scientists get onto the ice in a variety of ways. When the ship is stationary gangways can be put out and it is possible to simply walk on and off the ship. However, this is only true if the ship is going to remain at the site for some time. This is seldom the case and frequently sea-ice groups have a matter of only two or three hours (and even less than an hour at times) when the ship will stop for them to take samples. On these occasions they are often deployed onto the ice using cranes and metal baskets that hold two or three people with equipment. When ice is not thick enough to carry the weight of the workers and their often heavy equipment, sampling of the ice has to take place from within the baskets while they are suspended just above the ice floe surface.

This being said, it is remarkable how soon after freezing the ice becomes stable enough to stand on. Ice sheets of 10 cm (4 in) thick are more than enough to carry the weight of scientists and the diverse equipment they need. Even pancakes of ice just a few metres in diameter can support an ice team, although it can be a curious sensation swaying on a pancake of ice that is gently moving up and down in the waves, knowing that there is several kilometres of ocean beneath.

Sometimes the ship will not stop at all, and the sea-ice scientists are put onto the ice using helicopters. In this way a group can be deployed on the ice several hours ahead of the ship, so that when the scientists have finished their work they can be picked up either by helicopter again, or by the ship as it passes by.

Below
The *Polarstern* cutting through the pack ice.

Left and below
Scientists use the gangways put out from the ship (left), and are deployed on to the ice in metal baskets (below) to obtain samples.

Right
Antarctic field camp, including digging the freezer for storing food.

Opposite
Coring sea ice is the most common way to obtain ice samples.

Work on land-fast ice or coastal pack ice is, of course, far easier. If there is a suitable base or habitation nearby, access to the ice is far easier. It is simply a question of packing the equipment onto sledges and either pulling them or attaching them to motorised snow scooters or skidoos. This is by far the most convenient way of making sea-ice investigations, especially if the sites are close to well-equipped laboratories. An alternative is to set up temporary field camps on ice shelves or coastal land-fast ice regions that can be anything from a collection of tents to quite sophisticated living quarters and laboratories.

Coring the ice

Whether working from the ship or a land-based station, the main sampling method universally used for collecting ice samples is an ice corer. This is a barrel, typically 1 m (3 ft) long, that has cutting teeth on the bottom. By turning the shaft either by hand or using a small motor, it is possible to

Right
Taking a water sample from
under sea ice in the Baltic Sea.

sample cores of ice in 1 m (3 ft) sections. When the ice is thicker, extension pieces can be put onto the corer. In warm spring/summer ice, with air temperatures between 0 and –5°C (32 and 23°F), a 2–3 m (6½–10 ft) long core can take a matter of minutes to collect. In air temperatures around –30°C (–22°F) in February in the Fram Strait, a similar length core took four or five hours to collect.

The ice cores are often sub-sampled directly on the ice to prevent drainage of brine, which can alter the sample. Whether as sub-samples, or still as intact cores, the ice pieces are then returned to laboratories on the ship or a land base for further processing and analysis.

Divers and electrodes

One of the most difficult regions of the sea ice to sample is the fragile bottom layer. Traditional coring techniques may result in this part of the core being severely disrupted or lost altogether. The community living at the interface between the sea ice and the water below is rarely studied since this boundary is immediately disturbed upon any coring activity. The most obvious solution is to core or sample from beneath the sea ice using self-contained underwater breathing apparatus (SCUBA) and increasingly studies involve divers, although for obvious logistic and safety constraints these methods are best used in land-based studies where safety measures can be maximised. Disturbances from divers' exhaust air bubbles can also be problematic, often causing damage to the underside of the ice. Damage can be avoided by using surface-demand diving techniques or re-breathing equipment, where bubbles are not released.

One of the most exciting developments in oceanography over the past 15 years has been the development of fine electrodes that are capable of measuring the concentrations of pH, oxygen and carbon dioxide on a minute scale. Miniature light sensors are also available. These can be pushed into the ice from below and the photosynthesis and respiration dynamics recorded with great precision within individual brine channels. Surface ice is far too hard for such electrodes, and therefore access to the underside is vital to use these new technologies.

Once there is a hole in the ice there is good access to the underlying water. Therefore holes are often used to deploy equipment to collect water samples or measure profiles of salinity and temperature. In thick ice where coring is difficult, seal breathing holes can act as a convenient alternative to core holes.

Below
Using a Weddell seal hole to deploy water samplers under the ice.

Drift stations extend the work on the ice

One of the major frustrations of working in pack ice on a ship is the relatively short time there is for doing the work. Typically research cruises are between one and four months long. A sizeable percentage of each trip is spent steaming from the departure port to the working area, and so the actual time spent in the pack ice can be only a small fraction of the total cruise time, especially if there are other groups on board who want to work in open waters. To prolong the time spent working on the ice, there have been several drift stations or ice camps, where ships have been locked into the ice and allowed to drift with the pack ice for long periods.

There is a longer history of drift stations in the Arctic than in the Antarctic, and these are either set up by freezing a ship into the ice and building a camp around the ship, or deploying the equipment and camp personnel by air. Probably the most famous ice drift in ships were the pioneering drifts by Nansen on the *Fram* from 1900 to 1906, and Amundsen from 1918 to 1925 on the *Maud*. During the latter Finn Malmgren, a student on board the *Maud,* conducted some of the most fundamental sea-ice work.

Since these early ship drift experiments, there has been a large number of ice camps in various parts of the Arctic. The longest series of camps were those by Russian scientists who started annual pack-ice camps in 1937. The stations were generally constructed on multiyear ice floes about 2 km (1$^1/_4$ miles) in diameter and 3–5 m (10–16 ft) thick. The stations continued running until August 1991, only being interrupted by the two world wars. The series was started again in April 2003 with station NP-32, but ended dramatically when the 12 scientists had to be rescued in March 2004, when their ice floe broke up destroying most of the camp. These drift stations have collected some of the most comprehensive data sets available concerning weather patterns, ocean processes and, of course, sea ice.

One of the latest major ice camp drift experiments was Ice Station SHEBA, which started in the Arctic pack ice in October 1997 and finished in October 1998. The heart of the camp was the icebreaker *Des Groseilliers*, which was where the camp workers lived and communicated with the outside world. Around the ship a miniature village of containers, tents, plywood laboratories and measuring masts was established on an ice floe about 7.5 x 9 km (4$^1/_2$ x 5$^1/_2$ miles). About

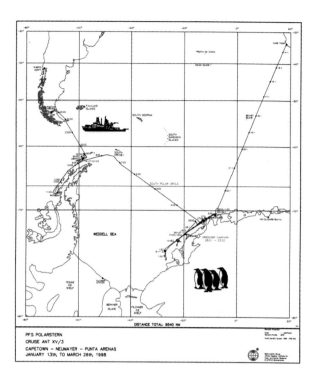

Above
Plot of a typical three-month Antarctic ship-based expedition with the research vessel *Polarstern*.

15 scientists worked at the camp during winter and about 35 in spring and summer, and crews and scientists were supplied and transferred using aircraft. In total 170 scientists used the station to conduct their work, which was fully supported by satellite imaging of the research area and from below with submarine investigations of sea-ice characteristics. During the year-long drift the floe travelled 1500 km (932 miles) ending up about 600 km (373 miles) northwest of where it started. The daily drift ranged from just a few hundred metres to over 30 km (19 miles).

Automatic drift stations

More recently researchers have conducted annual expeditions to the North Pole to establish automated measuring stations. The North Pole Environmental Observatories rely on the latest technology to record water salinity and temperature, ice thickness and the temperature of the ice at different depths as well as recording meteorological parameters. The fully automated arrays of equipment are deployed from airplanes in the spring. The equipment makes measurements and logs the data on internal computers while satellites track the position of the floe. These campaigns are supplemented by ice and oceanographic surveys, which are completed using helicopters and a series of underwater moorings.

Antarctic drift stations

Naturally the thick multiyear ice floes of the Arctic make such ice camps more feasible than in most of the thin pack ice that surrounds the Antarctic for relatively short periods of time. The main ice-drift stations to be attempted in the Antarctic pack ice have been confined to the western Weddell Sea, where the thickest ice is concentrated due to ice movements caused by the Weddell Gyre (p. 55). Of course, a forerunner of such a drift station, was the exploits of Shackleton and his crew once their ship, the *Endurance*, had sunk in the Weddell Sea. These men built camps of tents and hauled their equipment and lifeboats over the ice as they drifted with the ice on the Weddell Gyre. This is quite a different picture from the first fully international joint US-Russian Ice Station Weddell-1 (ISW-1) that drifted in the Weddell Sea from January to June 1992.

ISW-1 drew on the experiences of the long tradition of ice camps in the Arctic. Measurements were taken using instruments at the camp itself, as well as using helicopters and snowmobiles to collect data up to 150 km (93 miles) away from the camp. Two icebreakers, the *Nathaniel B. Palmer* and *Akademik Federov,* did not stay at the camp, but carried out complementary ship-based measurements in the region and were used to transport supplies and help transfer scientists to and from the camp, a task that was aided by aircraft as well. Over 30 scientists conducted work on physical oceanography, meteorology, sea-ice physics and biology.

A German-led ice-drift station (ISPOL) in the southwestern Weddell Sea will follow up this successful ISW-1 ice drift in 2004. The *Polarstern* will set off from Cape Town in November with the aim of drifting for 50 days anchored to a large ice floe. The 50 scientists and 50 crew on board will set up experiments on the ice, supported by two helicopters and teams of divers. Again a complex suite of scientific studies will be conducted from a camp consisting of fibreglass huts and tents. This will enable the researchers to have 24-hour access to their experiments and measuring devices deployed on and beneath the ice.

Airplanes, helicopters and submarines

These large-scale drift experiments are a great opportunity for sea-ice researchers, although the planning, logistical support and costs for such ventures are immense. Even with such experiments, a relatively small area of pack ice is actually covered. Besides the obvious advantages of satellite-based measurements (p. 47), aerial surveys using airplanes and helicopters can provide much valuable information on many aspects of sea-ice science, especially information on sea-ice thickness, surface roughness and the chemistry of the air above the ice and surrounding water.

Below
Setting up an ice camp in the Antarctic.

There have also been many studies conducted from submarines traversing the Arctic Ocean. Submarines have operated under Arctic sea ice since 1958, and these have been equipped with upwardly looking sonar devices that can be used to map the distribution of sea ice as well as the thickness of the ice. Many thousands of kilometres of profiles have been made, although it is only recently that a complete set of data has been released, since most of these activities were military operations. However, because the submarines travelled rather predictable transects on each journey, it is easy to compare inter-annual differences in sea-ice thickness and distribution for several decades over which these measurements were made.

Similar upward-looking sonar devices can also be attached to moorings attached to the sea floor. These sensors can be combined with others to record drift velocities and other oceanographic measurements. The sensors can be deployed in shallow or deep water and can be left for several years at a time recording the data on internal data loggers powered by batteries. In fact, it is the battery life that often determines how long a mooring can stay in the water, although there is always the chance that they will be lost because mooring wires are snagged by fishing gear or passing icebergs.

Ice tanks – sea-ice research in a laboratory

Sea-ice researchers are turning increasingly to laboratory simulations of sea-ice conditions because they allow the careful control of abiotic conditions governing sea-ice processes. In their most simple form, laboratory-simulation sea-ice experiments can be conducted in water volumes of the order of tens of litres, using the simple technology of a picnic cool box containing seawater being placed in a modified domestic freezer. A magnetic stirrer keeps the water moving. Such set ups can produce realistic sea-ice pancakes over three days.

Researchers have used facilities that normally test scale models of icebreaking ships and oilrigs in scaled-down pack-ice fields. These shallow 1.5 m (5 ft) deep basins – 20 m (66 ft) long and 6 m (20 ft) wide – can hold nearly 180 m³ (6356 ft³) of seawater and the air temperatures can be dropped to below –20°C (–4°F). Wave generators and pumps to create water currents are used to produce large areas of realistic sea ice, including the first pancake stages of ice. Scientists have used such basins to measure the biological changes in cultures of diatoms and bacteria as the water freezes and over long periods of ice growth. The ease with which physical and chemical conditions can be regulated in ice-tank facilities such as these is likely to take our understanding of sea-ice biology and physical processes much further, or certainly more swiftly, than would be possible by field campaigns alone. Another big advantage of tank facilities is the tremendous opportunity they offer for testing new sensors, sampling equipment and technologies for future use in the field.

Above
Taking samples to investigate the chemistry of sea ice produced in an ice-tank facility that simulates sea ice formed in the oceans.

Below left
Testing scale models of oil-rig evacuation vessels in an ice tank that simulates the pack-ice dynamics at the same scale as the models.

Pack ice: threats and potential

10

DESPITE THE INACCESSIBILITY OF THE POLAR REGIONS, INCREASINGLY MORE PEOPLE (TOURISTS, SCIENTISTS OR COMMERCIAL EXPLOITERS) ARE ABLE TO TRAVEL THERE. WITH INCREASED HUMAN ACTIVITY THERE IS A DANGER THAT THE PACK ICE OF THE ARCTIC AND SOUTHERN OCEAN MAY BE DAMAGED. POLLUTION, OIL SPILLS AND THE INTRODUCTION OF EXOTIC ORGANISMS ARE ALL OBVIOUS THREATS.

HOWEVER, THERE ARE several international treaties that aim to limit man's influence, and these regulations are rigorously controlled. The physiological and biochemical adaptations of bacteria, algae and animals in polar regions have attracted 'bioprospectors' seeking a source of novel pharmaceuticals, enzymes for use in industrial processes and other unusual compounds. Although we should encourage these biotechnological advances, it is important that the potential for exploitation is not to the detriment of what are still largely pristine environments.

Threats from man

It could be argued that the increasing numbers of tourists visiting polar regions are putting pressure on the very environment they so admire. Even though tourist trips remain hugely expensive, it can only be expected that the number of tourists wishing to visit the region will increase with time. However, the numbers of ships entering the pack-ice zones of the Arctic and Antarctic are very few indeed, and the tourist ships can only operate on the very edges of the marginal ice zones. Their objectives are normally land stations, and visits are only made during summer months.

There are strict international government regulations governing how close tourists can come to hauled out seals and penguin colonies, and strictly enforced regulations about the disposal of waste, and even the disinfection of footwear before stepping off the ship and on return. People who have conducted

Opposite
It is a major challenge to maintain the pack-ice regions of the world in a pristine condition.

research in the area often lead these trips, and are well placed to demonstrate the wonders of these regions, while at the same time causing minimum disturbance. All in all, although numbers of tourists will continue to rise, the environmental impact of the tourist industry on remote pack-ice regions of the world is probably negligible. Of course, careful monitoring of future developments needs to be enforced, but the ice is a very effective barrier for keeping people out of large areas of the ocean.

Below
Plastic rubbish washed up on an Antarctic shore with drifting pack ice.

Pollution issues

A far more sinister effect of mankind on the pack-ice covered waters of the world is pollution in its various guises. This can range from litter and discarded debris washing up on shores to heavy metal contamination, radioactive material and a diversity of organic pollutants, from PCBs to oil spills. One of the major concerns is that many of these pollutants get concentrated as they are passed up the food chain in a process called bioaccumulation. Of course, there are stringent regulations, especially concerning ships that operate in ice-covered seas. International protocols lay down strict guidelines on issues such as the strengthening of hulls to prevent oil and fuel spills, controlling the discharge of bilge water and sewage, and regulating exhaust emissions. Even the types of hull paint are controlled to prevent harmful chemicals entering polar waters.

Oil in the ice

Clearly people appreciate the risks from oil spills because of vivid images of oil spill disasters such as the *Exxon Valdez*. There are huge offshore oil developments in several areas of the Arctic Ocean, and in the Beaufort Sea alone more than 80 offshore exploratory wells have been drilled. Pipelines buried in the seabed largely bring the oil ashore. Naturally the establishment of such extensive oil operations result in thousands of survey, supply and construction boat traffic each year. These all have an impact on the environment, and there is an ever increasing fear of large-scale oil spills either from damage to the ships, or the oil platforms and pipelines in ice-covered waters.

It is common to think that oil spills are an Arctic problem and not pertinent to the Antarctic. In 1989, however, the *Bahia Paraiso* transporting fuel to a land station ran aground on the Antarctic Peninsula. It is estimated that 600,000 litres (132,000 gallons) of diesel fuel and oil leaked from the hull and four days later the oil slick covered an area of 30 km^2 (12 sq miles). There were obvious immediate impacts on the local bird populations, although the effects on other organisms were considered to be low, especially in the long term. The remaining fuel on the wreck was removed in 1992.

In open waters there are well-tested operations to contain and remove oil spills, but in ice-covered waters such spills are a very different proposition. The huge inflatable booms used to contain oil slicks tend to get caught by ice and even broken, and so containment is difficult. The major problem is that oil is not easy to get to, because instead of floating on the surface it will form slicks under the ice. This means that detergents or bacterial agents sometimes used to break down oil slicks cannot be used. If the oil spill happens during a phase of rapid ice growth, it is possible that the oil slicks may become incorporated into the ice floes and transported large distances by ice drift.

Oil-degrading bacteria can also grow within sea ice. In experiments, populations of bacteria have been shown to change from having no oil-degrading bacteria to 10% being comprised of these specialised bacteria after 30 weeks contamination with oil. There is a lot of interest in seeing whether or not these ice species have superior or novel mechanisms for breaking down diesel and other oils, compared to species of bacteria that break down oil in warmer waters.

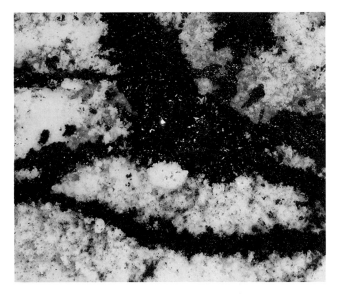

Cleaning up the act

Over the past 20 years town, village bases and camps on the shores of the Arctic and Antarctic have cleaned up their acts considerably concerning the types of material that they dump into the sea. Bases in Antarctica now have efficient sewage treatment works to prevent introducing organic matter and bacteria into coastal waters, although this is a relatively new development for some. Practically everything that is now taken to the bases has to be removed again after use. Therefore there are no more dumps of metal and organic waste, and a lot of time and money is being invested in cleaning up old bases and landfill

Top
Oil spills on sea ice may be broken down by bacteria within the ice.

Above
Ice tank experiments to see how spilt oil spreads over regions of pack ice.

sites and ensuring that contaminants do not find their way to the marine environment. It is paradoxical that the tourists, who many environmentalists fear may damage the fragile polar regions, are so restricted that they cause far less damage than the generations of scientists who preceded them.

Heavy metals, sediments and rivers

One of the major differences between the Antarctic and Arctic is the large rivers that discharge into the latter. Many of these rivers carry high quantities of heavy metals and organic pollutants into the shallow shelf regions of the Arctic. These can be incorporated into sea ice directly by being scavenged during frazil-ice formation or become buried in the sediments. Since Arctic sea ice can

Right
Before clean up (top), and the same site after removal of the waste (bottom).

Left
Pollutants can be
concentrated in the fat-rich
milk of seals that are then
passed on in high doses to
the young.

be heavily laden with sediments, polluting substances can become incorporated into sea ice by this means. Once in the ice the contaminants have the potential for being carried long distances into the open ocean. It is thought that in some parts of the coastal Arctic radioactive material may be contaminating sediments, and these may be being transported in sea ice. However, this level of contamination is arguably no greater than many other coastal regions where there is large industrial activity and big rivers. The difference is that sea ice may transport the material in a way, and to regions, that would not have been possible without the ice, such as in the trans-Arctic drift ice (p. 51).

It does seem that a whole class of organic pollutants are being concentrated in seals, whales and polar bears of the Arctic. These in turn are being concentrated in people eating those animals as part of their diet. These persistent organic pollutants (POPs) are often fat-soluble and therefore concentrated in the animals' fat reserves. They are also passed on from mothers to their young through milk that can be up to 60% fat. POPs have a range of effects from damaging the immune system to increasing the risk of tumours and reducing fertility.

Sea ice and biotechnology

In recent years there has been a lot of interest in understanding the ways in which organisms living within and at the edges of sea ice survive the freezing temperatures and high salinities of this habitat. The fact that bacteria remain active down to –20°C (–4°F) in winter Arctic sea ice means that their psychrophilic adaptations may be based on enzymes and membranes with unique characteristics that can be used for biotechnology and industrial processes. Sea-ice organisms are one of a large group of 'extremophiles' that are being isolated and screened for novel enzymes, metabolic capabilities and cellular modifications. Many of these organisms are thought to hold clues to our understanding of how life evolved on Earth. Sea-ice extremophiles have an increasingly important role in such studies alongside other organisms isolated from, among others, hot thermal springs, deep oceans and hydrothermal vents.

Cold-active enzymes

Enzymes from psychrophilic organisms have a variety of industrial uses. They are used in cheese maturation, polymer degradation and as additives in detergents and the production of pharmaceuticals. A number of cold-active enzymes have been identified and isolated from sea-ice bacteria and could help to reduce production costs. The main benefit of cold-active enzymes is that processes can be run with little or no external heat. Many industrial processes go through a sequence of reactions. Chemicals are often used to stop one stage to allow the next to start. By using cold-active enzymes, modest heat inactivation of the enzymes can be used rather than adding costly chemicals or filtration steps.

Diet supplements

One of the adaptations to ice temperatures is the production of PUFAs (polyunsaturated fatty acids, which prevent cell membranes from freezing), by ice bacteria and diatoms. Although healthy nutrition should provide all the nutrients we need, this does not stop the search for novel ways of enriching foodstuffs with substances such as PUFAs including the omega-3 fatty acids. More importantly PUFAs are necessary for the normal growth of the larvae and adults of many important aquaculture species, and like the krill (p. 119) they need to get them from their diet. There is much interest into finding ways of increasing PUFAs in animal feeds, and it is possible that strains of bacteria and diatoms from the ice may provide the key for bringing this about.

Astrobiology and snowball Earth

Sea ice and the organisms living within it are considered to be good analogues for life on ice-covered extraterrestrial systems, and also for previous periods in the Earth's history when ice caps covered most of the Earth's surface. There is convincing evidence that near the end of the Neoproterozoic era, between 1000 and 543 million years ago, ice up to a kilometre thick covered most of a 'snowball' Earth. Only a few mountains and volcanoes would be visible above the frozen surface. It is thought that psychrophilic organisms similar to those we know today from sea ice may have been able to survive these frozen times, when only the bottom of the oceans were not frozen as they are warmed from hydrothermal activity.

Moving on in Earth's history there were major ice ages about 430–445 million years ago during the Ordovician and Silurian times, and then 260–340 million years ago during the Carboniferous and Permian periods. A little more recently, at the height of the last ice age 21,000 years ago, much of North America and Europe were covered in huge glaciers, and land and sea ice covered 30% of the Earth's surface. Again psychrophilic organisms, especially those able to survive in the ice, would have been one of the major groups of organisms to survive the frozen oceans, especially where ice reached equatorial latitudes and was presumably thinner than at higher latitudes, and therefore light could penetrate the ice and photosynthesis could take place.

And on to Mars, Europa and beyond?

It appears that there have also been ice ages on Mars, and many geological features on Mars closely resemble the remnants of glaciers and other ice features. There is also evidence that water still exists on the planet.

The *Voyager* and *Galileo* space expeditions have shown that Europa, a moon of Jupiter, is covered with brown and white ice, overlying a briny ocean. It is estimated that this frozen surface is a few kilometres thick and may even be up to 100 km (62 miles) thick. The cracked surface of the ice on Europa is thought to be due to tidal stresses.

As soon as the first images of brown cracked ice on the surface of Europa were studied, it wasn't long before scientists made a link between diatoms that colour sea ice brown and these fascinating structures. There is a tantalising prospect to compare the two systems, and more evidence is being produced to show that bacteria can survive some of the pressures or forms of ice that may exist on Europa. Ganymede and Callisto, two other moons of Jupiter, are also thought to have ice on their surfaces and so may be potentially life supporting.

Left
The surface of Europa, a moon of Jupiter, is covered by ice floating on top of a salty ocean.

Right
Not all equipment needs
to be sophisticated. Flags are
used to mark safe routes,
experimental sites and
storage areas. Without
them we would be lost.

Whether there are life forms on Europa or Mars, or have been in the past, the sea ice on Earth provides scientists with a suitable analogue. By studying bacteria that survive in cold, thick, Arctic ice, where almost no water is left surrounding the cells, we may learn much about potential life forms. When expeditions to sample these environments are devised, it is certain that a lot of the testing of equipment and sampling strategies will be based on the experiences gained from working within the pack ice, in particular thick deformed multiyear ice that is most typical of the central Arctic.

A hundred years on

The pack-ice regions of the world are still places that only few venture to, and vast parts of them have never been visited by any human at all. They cover unexplored oceans, and form an effective barrier that still defeats much of our engineering and technological prowess. Despite man venturing into the 'Mare concretum' many hundreds of years ago, it is only in the past 100 years that we have really begun to understand the processes governing the pack ice and the organisms that live within, on top, and beneath the frozen skin. In fact most of our understanding is based on just the last 50 years of endeavour. There are many exciting scientific discoveries that have already been made and many still await us, some of which may take us to the moons of Jupiter no less. It is clear we need to understand the pack ice to be able to understand changing climate and large-scale ocean processes. But despite all of these pertinent quests, at the end of the day it is the raw, hostile beauty that is one of the greatest lures to draw people to the pack ice.

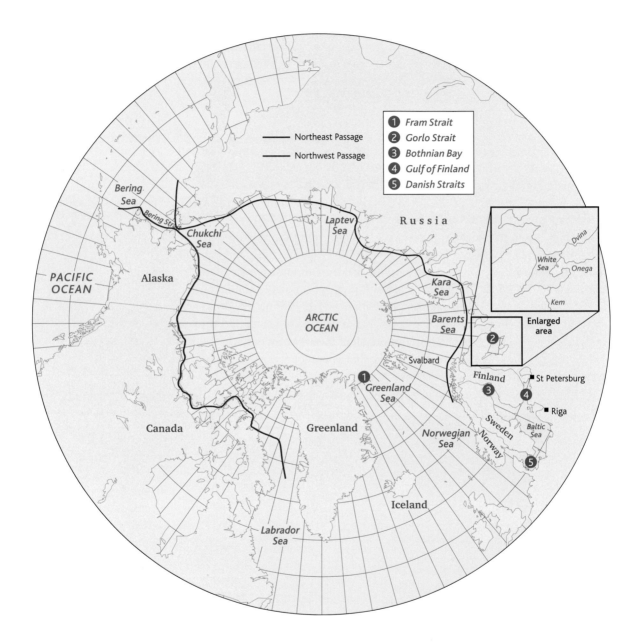

Locations in the Arctic
Ocean, White and Baltic
Seas, referred to in the text.

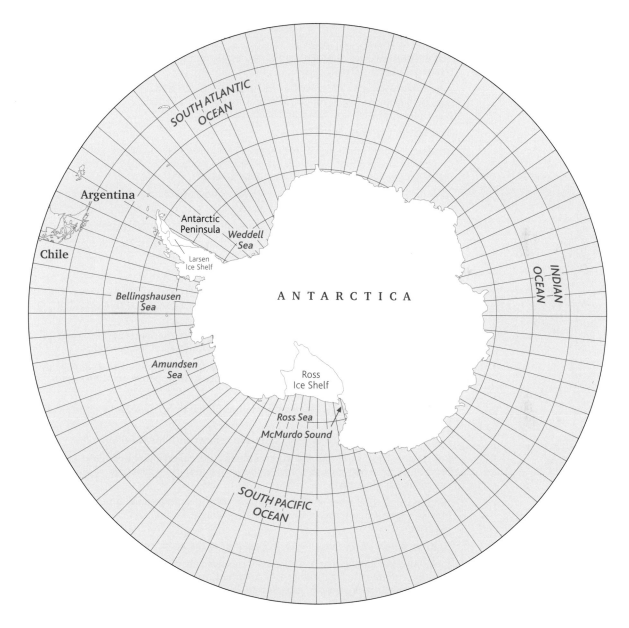

Map to show the
Southern Ocean
surrounding Antarctica.

Glossary

Abiotic: Non-biological, describing factors such as atmospheric gases, inorganic salts and mineral particles. It also describes chemical and physical factors such as temperature and salinity.

Accessory pigments: Photosynthetic pigments of plants and algae that trap light energy in addition to chlorophyll a. Accessory pigments include the carotenoids, fucoxanthin, phycobiliproteins, and chlorophylls b, c, and d.

Acoel: Lack of a coelom (body cavity).

Aerobic: Aerobic micro-organisms (aerobes) are those that require oxygen for growth; obligate aerobes cannot survive in the absence of oxygen. The opposite are anaerobic organisms, which do not require oxygen for growth.

Albedo: The fraction of the incident short-wave radiation that is reflected from a surface.

Algae: General term for eukaryotic, chlorophyll-containing, oxygen-producing photosynthetic organisms that are not plants in a strict taxonomic sense.

Amphipods: A member of an order of crustaceans, Amphipoda. They are generally flattened laterally, sometimes even swimming on their sides.

Anaerobic: See **Aerobic**

Antarctic Circumpolar Wave (ACW): A propagating wave observed to cause disruptions at the sea-ice margin. It moves around the Antarctic with a period of approximately eight years.

Antifreeze proteins: Proteins produced by an organism that prevent freezing of its tissues or body fluids when subject to subzero environmental temperatures.

Archaea: Prokaryotic organisms belonging to the third major domain of life (the other two are the Eubacteria and Eukaryota), with most known species associated with extreme environments.

Baleen whale: Large cetacean, whose mouth is lined by bristles used to strain food from seawater.

Benthic: Refers to the sea floor, or waters in contact with the sea floor.

Benthos: In freshwater and marine ecosystems, the collection of organisms attached to or resting on the bottom sediments (epifaunal), and those which bore or burrow into the sediments (infaunal).

Biomass: The total mass of all the organisms of a given type and/or in a given area. May be expressed in terms of wet weight, dry weight, carbon and chlorophyll, among many others.

Bottom water: Water mass at the deepest part of the water column in the ocean.

Carnivore: An animal that feeds largely or exclusively on other animals. The term is used most frequently for members of the order Carnivora, which includes flesh-eating mammals such as dogs, cats, bears and seals.

Chlorophyll: A group of green pigments in photosynthetic organisms involved in harvesting light by absorption, excitation and transfer of energy.

Ciliates: Protozoa, up to 0.15 mm long, which swim actively, engulfing small food particles via the mouth and gullet.

Copepods: A small aquatic crustacean of the class Copepoda, without a carapace and with paddle-like feet. Many are minute and form a major component of marine plankton.

Detritus: Dead particulate organic material.

Diatom: Unicellular algae with a cell wall or frustule of silica. Adapted to a wide range of pelagic and benthic marine habitats.

Dimethylsulphide (DMS): A volatile sulphur compound produced by marine phytoplankton and transferred to the atmosphere. Following oxidation to sulphate particles, DMS affects the radiative properties of the atmosphere by reflecting solar radiation and by affecting the concentration of cloud condensation nuclei (CCN). It is the main natural source of sulphate aerosol and the major route by which sulphur is recycled from the ocean to the continents. The production of dimethylsulphoniopropionate (DMSP), the precursor of DMS in the ocean is strongly dependent on the phytoplankton species.

Dimethylsulphoniopropionate: See **Dimethylsulphide**

Dinoflagellate: A single-celled micro-organism capable of propulsion through use of whip-like flagella. Can be covered in armoured cellulose plates or naked.

Dispersal: Moving from an area of origin or growth to another area.

ENSO cycle: El Niño Southern Oscillation, which refers to the coherent, large-scale fluctuation of ocean temperatures, rainfall, atmospheric circulation, vertical motion and air pressure across the tropical Pacific.

Eukaryotes: Organisms with a cellular nucleus and typically having DNA arranged in multiple chromosomes.

Euphausiids: 1–7 cm ($^1/_2$–3 in) shrimp-like crustaceans of the phylum Arthropoda.

Extremophile: A micro-organism that thrives under extreme environmental conditions of temperature, pH or salinity.

Fast ice: Permanent, often multiyear ice covering oceanic/fjord water which is literally stuck 'fast' to the continent or to some other fixed objects such islands, grounded icebergs, and peninsulas. Often referred to as land-fast ice.

Fatty acid: An organic acid with a long straight hydrocarbon chain and an even number of carbon atoms. Fatty acids are the fundamental constituents of many important lipids, including triaglycerides. In animals some fatty acids can be synthesised by the body, others such as polyunsaturated fatty acids must be obtained from the diet.

Foraminifer: A protozoan of the mainly marine order Foraminifera, having a perforated shell through which amoeba-like pseudopodia emerge. When they die the shells eventually drop to the seabed to form foraminiferal ooze.

Frazil ice: Fine spicules or plates of ice suspended in water during ice formation.

Freeboard: The height of the ice surface above the water level.

Frustule: Silica cell wall of a diatom, generally two per cell.

Grazing: The consumption of algae by herbivores including protozoa and zooplankton.

Grease ice: A soupy thin layer of ice at the water surface formed from the accumulation of frazil ice.

Harpacticoid: Group of copepods which live on the seabed, which are non-planktonic.

Herbivore: An animal that feeds on plants or algae.

Heterotrophic: The ability to obtain carbon for organic synthesis by metabolising organic materials.

Ice area: The actual area (extent) covered by ice.

Ice concentration: The percentage of ice-covered surfaces within the satellite footprint or grid. Often expressed as a number out of ten, i.e. a scale from 1/10 to 10/10.

Ice edge: The demarcation between ice-free ocean (or sea) and ice-covered ocean (or sea) and is usually set at 15% ice concentration when using satellite microwave data.

Ice extent: The total sum of the area covered by sea ice of at least 15% sea-ice concentration.

Ice floe: A discrete section of sea ice that varies in area from a few square metres to a few thousand square metres.

Interstitial: The space between adjacent particles in a substrate, or ice crystals in platelet layers.

Krill: Norwegian word meaning whale food, and usually referring to schooling crustaceans of the family Euphausiacea. Often specifically *Euphausia superba* in the Southern Ocean.

Lead: A linear feature of open water that occurs in the ice pack between ice floes caused by ice break-up, and may be covered by new ice.

Lysis: Rupturing of a cell and loss of cell contents.

Marginal ice zone: (MIZ) A geographical zone at the ice edge where ice floes are relatively loose and influenced by waves.

Metazoans: Multicellular animals. c.f. protozoa.

Mixotrophic: The ability to obtain carbon for organic synthesis from both inorganic sources (autotrophy) and organic materials (heterotrophy).

Multiyear ice: Sea ice that has survived at least one summer melt period.

Mycosporine-like amino acid (MAA): A molecule that absorbs strongly in the UV range. Many organisms produce this molecule for use as a sunscreen.

Nauplii: Early developmental stage in many crustaceans such as copepods and euphausiids.

Nematodes: (Nematoda) Small cylindrical roundworms of the phylum Aschelminthes.

New ice: A general term for newly formed ice, which includes frazil ice, grease ice and nilas.

Pack ice: All sea ice except that which is attached to land. See 'fast ice'.

Pancakes: Predominantly circular pieces of ice from about 3 cm to 3 m (1¼ in to 10 ft) in diameter, less than 10 cm (4 in) in thickness and with raised edges.

Pelagic: Describing organisms that swim or drift in the open water of a sea or a lake, as distinct from those that live on the bottom (see benthos).

Phospholipids: A group of lipids that consist of a phosphate group and one or more fatty acids. They are major components of cell membranes in plants and animals.

Photosynthesis: The process by which the energy of sunlight is used by green plants and algae and some bacteria to synthesise carbohydrates from carbon dioxide and water.

Photosynthetically available radiation (PAR): Radiation at wavelengths of 400–700 nm (visible light), which are absorbed by algae and plants for photosynthesis.

Phytoplankton: The photosynthesising organisms of plankton (see below), consisting chiefly of microscopic algae, such as diatoms and dinoflagellates.

Plankton: Pelagic organisms (mostly microscopic) that drift or float passively with the current in a sea or lake. Plankton includes many microscopic organisms, such as algae, protozoa, various animal larvae, and some worms. It forms an important food source for many other members of the aquatic community and is divided into zooplankton and phytoplankton.

Platelet ice: Leaf or disc like ice crystals either free floating or fixed under sea ice, usually fast ice.

Polychaetes: (Polychaeta) Marine segmented worms of the phylum Annelida, some are planktonic, but most are benthic, many with meroplanktonic larvae.

Polynya: A large irregular opening of water enclosed by ice. May be formed near coastal areas by wind displacement, or by the upwelling of warm water.

Polysaccharide: Any of a group of carbohydrates comprising long chains of monosaccharide (simple sugar) molecules.

Polyunsaturated fatty acids (PUFA): Fatty acids with two or more double bonds. Organic acids consisting of carbon chains with a carboxyl group at the end.

Pressure ridge: An elongated ridge or wall of broken ice forced up by ice pressure between two ice floes.

Primary production: The photosynthetic carbon fixation per unit area, per unit of time.

Prokaryotes: Organisms lacking a cellular nucleus and typically having DNA in a single molecule.

Protist: General term for eukaryotic microbes (algae and protozoa).

Protozoa: Mostly microscopic unicellular organisms that graze on bacteria and phytoplankton.

Psychrophilic: Describing an organism that lives and grows optimally at relatively low temperatures, usually below 15°C (59°F) and cannot grow above 20°C (68°F). Psychrophiles, which consist mainly of bacteria, algae, fungi and protozoa, are restricted to permanently cold climates.

Psychrotolerant: Organisms live and grow best at temperatures of 20–40°C (68–104°F), but are able to tolerate cold conditions.

Remineralise: Transforming organic materials into inorganic forms, such as organic carbon to CO_2.

Remote sensing: The gathering and recording of information concerning the Earth's surface by techniques that do not involve actual contact with the object or area under study. These techniques include photography (e.g. aerial photography), multispectral imagery, infra-red imagery and radar. Remote sensing is generally carried out from aircraft and, increasingly using satellites.

Salinity: The number of grams of salts dissolved in 1000 g of water.

Salps: Barrel-shaped planktonic tunicates of the phylum Chordata. Salps are filter-feeders.

Sea ice concentration: The fraction of open water within sea-ice cover. For example, 0–15% = open ocean, 100% = total ice cover.

Strand community: An ice algal community that extends vertically into the upper water column from an ice floe.

Texture: Typically describes the spatial arrangement and orientation of sea-ice crystals.

Thermohaline: The combined impact of temperature and salinity on seawater density. In the polar oceans this is dominated typically by the impact of salinity, in particular in such areas where the loss of highly saline brine from growing sea ice can substantially increase seawater density and induce convective overturning and amplify thermohaline mixing.

Triacylglycerol: The major constituent of fats and oils that provide a concentrated food energy store in living organisms.

Trophic: Used to describe feeding relationships, such as levels in a food chain.

Turbellarians: (Turbellaria) oval to elongated unsegmented flatworms of the phylum Platyhelminthes.

Zooplankton: Animals of the plankton. See plankton.

Index

Further information

Books

Pack-ice research is covered in a wide range of disciplines and is written about from many different angles. The following list is only a small fraction of fairly recent titles that may form the basis of further reading about pack ice.

A History of Antarctic Science, G.E. Fogg. Cambridge University Press, 1992.

Antarctica: A Guide to the Wildlife, Tony Soper and Dafila Scott. Bradt Travel Guides, 2000.

Antarctica: The Complete Story, D. McGonigal and L. Woodworth. Frances Lincoln, 2003.

Antarctic Microbiology, E.I. Friedman (ed). Wiley-Liss, 1993.

Endurance: The Greatest Adventure Story Ever Told, A. Lansing. First published 1959.

Endurance: Shackleton's Incredible Voyage to the Antarctic. New ed. Weidenfeld & Nicholson, 2001. Illustrated ed. Weidenfeld & Nicholson, 2001.

Exploring Polar Frontiers: A Historical Encyclopedia (2 volumes), W.J. Mills. ABC-CLIO, 2003.

Farthest North, F. Nansen. Abridged ed. Gerald Duckworth & Co. Ltd, 2000.

H_2O: A Biography of Water, P. Ball. Weidenfeld & Nicholson, 1999.

Ice Drift, Ocean Circulation and Climate Change, J. Bischoff. Springer-Praxis, 2000.

Ice in the Ocean, P. Wadhams. Gordon & Breach Science Publishers, 2000.

Illustrated Glossary of Snow and Ice, T. Armstrong, B. Roberts and C. Swithinbank. Scott Polar Research Institute, Special Pub. 4 (2nd edn.). The Scholar Press, 1973.

Krill: Biology, Ecology and Fisheries, I. Everson. Blackwell Science, 2000.

Life in the Freezer, A. Fothergill. BBC Worldwide, 1993.

Lonely Planet: Antartica (Country and Regional Guides), Jeff Rubin. 2nd ed. Lonely Planet Publications, 2000.

Lonely Planet: Antarctica: A Travel Survival Kit (Travel Survival Kits), Jeff Rubin. Lonely Planet Publications, 1996.

Lonely Planet: Arctic (Country and Regional Guides), Deanna Swaney. Lonely Planet Publications, 1999.

Penguins: Living in Two Worlds, L. Davis. Yale University Press, 2004.

Physics of Ice Covered Seas, M. Leppäranta (ed). Helsinki University Press, 2001.

Polar Bears, I. Stirling. The University of Michigin Press, 1998.

Sea ice: An Introduction to its Physics, Chemistry, Biology and Geology, D.N. Thomas and G.S. Dieckmann (eds). Blackwell Publishing, 2003.

Search for Life, M. Grady. The Natural History Museum, 2001.

The Adélie Penguin: Bellweather of Climate Change, D.G. Ainley. Columbia University Press, 2002.

The Arctic: A Guide to Coastal Wildlife, Tony Soper and Dan Powell. Bradt Travel Guides, 2001.

The Biology of Polar Habitats, G.E. Fogg. Oxford University Press, 1998.

The Biology of the Southern Ocean. G.A. Knox. Cambridge University Press, 1994.

The Blue Planet, A. Byatt, A. Fothergill and M. Holmes. BBC Worldwide, 2001.

The Endurance: Shackleton's Legendary Antarctic Expedition, C. Alexander. Bloomsbury Publishing, 1998.

The Great White South, H.G. Ponting. Cooper Square Press, 2002.

The Ice, S.J. Pyne. Weidenfeld & Nicholson, 2003.

The Oceans and Climate, G. Bigg. Cambridge University Press, 2003.

The South Pole: An account of the Norwegian Antarctic Expedition in the Fram 1910–1912, R. Amundsen. New York University Press, 2001.

The Voyage of Discovery: Scott's First Antarctic Expedition, R.F. Scott. Cooper Square Press, 2001.

Ultima Thula: Arctic Explorations, M. Lainema and J. Nurminen. John Nurminen Foundation, 2001.

Specialised journals

There are several specialist journals that cover aspects of Arctic and Antarctic research, including the latest research into pack-ice studies. However, since the implications of pack-ice research are so widespread, much of the research is reported in more general journals, a few of which are also included in the following list:

Antarctic Science, Cambridge University Press

Arctic, Journal of the Arctic Institute of North America

Cold Regions Science and Technology, Elsevier Publishing

Extremophiles, Springer-Verlag

Journal of Geophysical Research and *Antarctic Research Series*, American Geophysical Union

Journal of Glaciology and *Annals of Glaciology*, International Glaciology Society

Nature, MacMillan Publishing

Polar Biology, Springer-Verlag

Polar Record, Cambridge University Press

Science, American Association for the Advancement of Science

Websites

There are many websites that deal with sea-ice related issues. However, these are subject to change, and it seems prudent to simply list some of the organisations that work with sea ice, and for readers to follow links to the specialised sea-ice material. This is not intended to be an exhaustive list, but rather the first stage to help access information related to sea ice on the web. Much of the research into pack ice is undertaken by researchers in universities. These have not been included in this list, although many of the following institutions have links to their university colleagues on their web pages.

Alfred Wegener Institute for Polar and Marine Research – http://www.awi-bremerhaven.de

Arctic and Antarctic Institute, Russia – http://www.aari.nw.ru

Arctic Council – http://www.arctic-council.org

Arctic Institute of North America – http://www.ucalgary.ca/AINA

Australian Antarctic Division – http://www.aad.gov.au

Antarctic Cooperative Research Centre, Australia – http://www.antcrc.utas.edu.au/antcrc

British Antarctic Survey – http://www.antarctica.ac.uk

Canadian Ice Service – http://ice-glaces.ec.gc.ca

Canadian Polar Commission – http://www.polarcom.gc.ca

Canadian Wildlife Service – http://www.cws-scf.ec.gc.ca/index_e.cfm

Cold Regions Research and Engineering Laboratory – http://www.crrel.usace.army.mil

Danish Polar Center – http://www.dpc.dk

Finnish Institute of Marine Research – http://www.fimr.fi

Hamburg Ship Model Basin (HSVA) – http://www.hsva.de

International Arctic Research Center, Alaska – http://www.iarc.uaf.edu

Inuit Circumpolar Conference – http://www.inuitcircumpolar.com

NASA – http://www.nasa.gov

NASA Earth Observatory – http://earthobservatory.nasa.gov

NASA, JPL Oceanography Group – http://oceans-www.jpl.nasa.gov/polar

National Oceanic and Atmospheric Administration – http://www.noaa.gov

National Institute of Polar Research, Japan – http://www.nipr.ac.jp

National Snow and Ice Data Center – http://nsidc.org

New Zealand Antarctic Institute; – http://www.antarcticanz.govt.nz

Norwegian Polar Institute – http://npiweb.npolar.no

SeaWifs project – http://seawifs.gsfc.nasa.gov/SEAWIFS.html

Scientific Committee on Antarctic Research – http://www.scar.org

Scott Polar Research Institute – http://www.spri.cam.ac.uk

The Astrobiology Web – http://www.astrobiology.com

Picture credits

Front cover ©David N. Thomas; pp.5-8 ©David N. Thomas; p.9 ©MODIS/D. Vaughan/BAS; p.11(t) ©David N. Thomas, (b) ©Cornelia Thomas; p.13 ©NASA GSFC Image by Reto Stöckli, Robert Simmon/MODIS teams; pp.14-19 ©David N. Thomas; p.20(l) ©Courtesy K.-U. Evers (HSVA), (r) ©Christian Haas; pp.21-22 ©David N. Thomas; p.23 ©Jurgen Weissenberger, Gerhard Dieckmann, AWI for Polar and Marine Research; pp.24-27 ©David N. Thomas; p.28 (t) Josefino Comiso NASA GSFC, (b) amended from original by Christian Haas, from *Sea Ice*, D. N. Thomas *et al.*, ©Blackwell Publishing; pp.29-34 ©David N. Thomas; p.35 ©Hajo Eicken; pp.37-41 ©David N. Thomas; p.42 Mike Eaton/©NHM (adapted from an Open University publication); p.44 ©David N. Thomas; p.46 Josefino Comiso NASA GSFC, from *Sea Ice*, D. N. Thomas *et al.*, ©Blackwell Publishing; p.47 ©Christian Haas; p.48 Josefino Comiso NASA GSFC, ©American Geophysical Union; p.49(t) amended from original ©Josefino Comiso NASA GSFC, (b) ©Courtesy of JAXA; p.50 amended from original ©Peter Wadhams; p.51 Josefino Comiso NASA GSFC, ©American Geophysical Union; p.53 amended from original ©Jim Hurrell; p.54 ©Courtesy of JAXA; p.55 amended from original ©Michael P. Schodlok (AWI); p.56 ©David N. Thomas; p.57 amended from original ©Michael P. Schodlok (AWI); p.58 ©David N. Thomas; p.60 amended from original by Commonwealth Bureau of Meteorology (Australia) source http://bom.gov.au/climate; p.63 ©MODIS Rapid Response Team at NASA GSFC; pp.64-65 ©David N. Thomas; pp.66-67 ©MODIS Rapid Response Team at NASA GSFC; pp.68-69 ©David N. Thomas; p.70(t) ©NHM, (tm) ©Gerhard Dieckmann, (bm), (b) ©James A. Raymond; p.71(t) ©Sönnke Grossmann, ©NHM; p.72(t) amended from original by Sigi Schiel from *Sea Ice*, D.N. Thomas *et al.*, ©Blackwell Publishing, (b) ©Iris Werner; p.73 ©J. Plötz, AWI for Polar and Marine Research; p.74 ©David N. Thomas; p.76 Karen Junge ©Reprinted from *Annals of Glaciology* with permission from the International Glaciological Society; p.78(t) ©David N. Thomas, (b) amended from original by Kevin Arrigo, from *Sea Ice*, D. N. Thomas *et al.*, ©Blackwell Publishing; p.79 ©Christian Haas; p.80 ©David N. Thomas; p.81 ©Cornelius Sullivan; p.82 ©Gerhard Dieckmann, p.83 ©Rod Budd/National Institute of Water and Atmospheric Research Ltd, NZ; p.84 ©David N. Thomas; p.85 ©Mark Brandon; p.86 ©David N. Thomas; p.88 ©Jan Michels AWI for Polar and Marine Research; p.89 ©James Raymond; pp.91-93 ©David N. Thomas; p.95 ©Julian Gutt; p.96(t) ©Thomas Mock, (b) ©David N. Thomas; p.97 ©David N. Thomas; p.98 ©NASA; p.100 ©Orbital Imaging Corporation and processing by NASA GSFC; p.103 ©Marcia M. Gowing; p.104(t) ©Borris, M., Helmke E., Hantschke, R. and Schweder, T. *Extremophiles* (2003) Vol 7: 377-384; p.104(tm) and (bm) ©Brent Christner, (b) ©Seija Hallfors/Finnish Institute of Marine Research; p.106(t) amended from original ©David N. Thomas, (b) Karen Junge ©Reprinted from *Annals of Glaciology* with permission from the International Glaciological Society; pp.107-109 ©Anna Pieńkowski-Furze; p.110(t) ©Wolfgang Petz, (b) ©Jeremy Young/NHM; p.111 ©Gerhard Dieckmann; pp.112-115 ©BAS; p.116 ©Iris Werner; p.117 ©BAS; p.118 ©P. Marschall/AWI for Polar and Marine Research; p.119 redrawn from original ©BAS; p.120 Reprinted with permission from *Science*, Vol 295, Issue 5561, 1890-1892 , March 2002, Brierley *et al.*, Antarctic krill under sea ice: elevated abundance in a narrow band just south of the ice edge ©2004 AAAS; p.121 amended from original ©Andrew Brierley; pp.123-125 ©Christine Friedrich; p.126 ©J. Plötz, AWI for Polar and Marine Research; p.127 ©Dr Tony North; p.128 ©Wolfgang Petz; p.130 ©Jenny Beaumont/BAS; p.132 ©Chris Gotschalk; pp.133-136 ©Julian Gutt; (b) ©Rod Budd/National Institute of Water and Atmospheric Research Ltd, NZ; pp.139-140 ©Julian Gutt; p.141 ©Dylan Evans; p.142 ©Gerhard Dieckmann; pp.143-146 ©David N. Thomas; p.148(t) ©Kirsty Brown c/o David Barnes, (b) ©Julian Gutt; p.150 ©David N. Thomas; p.152 table amended from *Sea Ice*, D.N. Thomas *et al.*, ©Blackwell Publishing; p.153, ©154(t) ©Bryan & Cherry Alexander Photography; p.154(b), p.155 ©David N. Thomas; p.156 ©Lee A. Fuiman; pp.157-158 ©David N. Thomas; pp.160-162 ©Bryan & Cherry Alexander Photography; p.163 ©J. Plötz, AWI for Polar and Marine Research; p.164(t) ©Sue Flood/NaturePL, (b) ©Mark Brandon; p.165 ©Frank Todd/B&C Alexander; p.166 ©David N. Thomas; p.167 ©Lydersen *et al.* 2003 *Geophysical Research Letters*, Vol. 29, Salinity and temperature structure if a freezing Arctic fjord-monitored by white whales; p.169 ©David N. Thomas; p.170 ©J. Plötz, AWI for Polar and Marine Research; pp.171-172 ©David N. Thomas; p.173 ©J. Plötz, AWI for Polar and Marine Research; p.174(t) ©Bryan & Cherry Alexander Photography, (b) ©Frank Todd/B&C Alexander; p.176 ©David N. Thomas; p.178 amended from original ©Maps.com; p.179 amended from original *Arctic Pollution*, ©Arctic Monitoring and Assessment Programme (AMAP), Oslo, Norway, 2002, xii+112 p.; pp.181-184 ©NHM; p.185 ©Scott Polar Research Institute; pp.186-188 ©David N. Thomas; p.189 (t) ©David N. Thomas, (b) ©Cornelia Thomas; pp.190-192 ©David N. Thomas; p.193 ©E. Nothig; pp.194-198, p.199(t) ©David N. Thomas; p.199(b) ©Courtesy K.-U. Evers (HSVA); p.200 ©David N. Thomas; p.202 ©David Barnes, BAS; p.203(t) ©Gerhard Dieckmann, (b) ©Courtesy K.-U. Evers (HSVA); p.204 ©BAS; p.205 ©Bryan & Cherry Alexander Photography; p.207 ©NASA/JPL/Caltech; p.209 ©David N. Thomas; p.210-211 Lisa Wilson/©NHM; back cover ©David N. Thomas.

Every effort has been made to contact and accurately credit all copyright holders. If we have been unsuccessful, we apologise and welcome corrections for future editions or reprints.

AWI, Alfred Wegener Institute; BAS, British Antarctic Survey;
GSFC, Goddard Space Flight Center; NHM, Natural History Museum, London.

Author's acknowledgements

I am grateful to Gerhard Dieckmann, Hajo Eicken, Christian Haas, Sigi Schiel, Gerhard Kattner, Stathis Papadimitriou, Markus Gleitz, Sönnke Grossmann, Victor Smetacek, David Roberts, Tony Fogg, Celia Coyne, Jean-Louis Tison, Hilary Kennedy, Christopher Krembs, Joey Comiso, Kevin Arrigo, Michael Lizotte, Andy Brierley, Thomas Mock, Andreas Krell, Christian Wiencke, Gunter Kirst, Ulf Karsten, Gaby Weykam, Christina Langreder, Horst Bornemann, Joachim Plötz, Mark Brandon, Finlo Cottier, Peter Wadhams, David Crane, Hartmut Hellmer, Andrei Naumov, Mats Granskog, Hermanni Kaartokallio, Harri Kuosa, Jurgen Weissenberger, Jörg Bareiss, Andy Clarke, Julian Priddle, David Ainley, Cynthia Tynan, Ian Stirling, Amy Leventer, Leanne Armand, Rolf Gradinger, Kalle Evers, Sven Günther, Virginia Giannelli, Rubén Lara, Peter Williams, but most of all Cornelia Thomas, for their help in various ways in the realisation of this project.

Since 1996 the following institutions have supported my sea-ice research: The European Union, The Nuffield Foundation, The Leverhulme Trust, The British Council, The Royal Society, The Finnish Institute for Marine Research and The Natural Environment Research Council. This work was produced as a result of opportunities offered by the University of Wales, Bangor, The Hanse Institute for Advanced Study, Germany, and The Alfred Wegener Institute for Polar and Marine Research, Germany.